Primary LENTIL SCIENCE

Pour the same amount.
See how high the lentils rise!

Science with Simple Things

Conceived and written
by RON MARSON

Edited and illustrated
by PEG MARSON

TOPS LEARNING SYSTEMS
10970 South Mulino Road
Canby OR 97013

Online catalog: www.topscience.org
Business office: 503.266.8550
Toll free (for orders only, please): 888.773.9755

Oh, those pesky COPYRIGHT RESTRICTIONS !

Dear Educator,

TOPS is a nonprofit organization dedicated to educational ideals, not our bottom line. We have invested years of love and hard work, and many tens of thousands of dollars, to bring you this excellent teaching resource.

And with this book we bring you unprecedented economy. Each lesson is folded like a card, with student instructions and materials outside, teaching notes at your fingertips inside. You need only reproduce them once and students can use them over and over, year after year. If you consider the depth and quality of this curriculum amortized over years of teaching, it is dirt cheap, orders of magnitude less than prepackaged kits and textbooks.

Please honor our copyright restrictions. We are a very small company, and book sales are our life blood. When you buy this book and use it for your own teaching, you sustain our publishing effort. If you give or "loan" this book or copies of our Job Cards to other teachers, with no compensation to TOPS, you squeeze us financially, and may drive us out of business. Our well-being rests in your hands.

What if you are excited about the terrific ideas in this book, and want to share them with your colleagues? What if the teacher down the hall, or your homeschooling neighbor, is begging you for good science, quick! We have suggestions. Please see our *Purchase and Royalty Options* below.

We are grateful for the work you are doing to help shape tomorrow. We are honored that you are making TOPS a part of your teaching effort. Thank you for your good will and kind support.

Sincerely,

Ron Marson

Purchase and Royalty Options:

Individual teachers, homeschoolers, libraries:

PURCHASE option: If your colleagues are asking to borrow your book, please ask them to read this copyright page, and to contact TOPS for our current catalog so they can purchase their own book. We also have an **online catalog.** See our addresses on page 1.

If you are reselling a **used book** to another classroom teacher or homeschooler, please be aware that this still affects us by eliminating a potential book sale. We do not push "newer and better" editions to encourage consumerism. So we ask seller or purchaser (or both!) to acknowledge the ongoing value of this book by sending a contribution to support our continued work. Let your conscience be your guide.

Honor System ROYALTIES: If you wish to make copies from a library, or pass on copies of just a few activities in this book, please calculate their value at 50 cents (25 cents for homeschoolers) per Job Card per recipient. Send that amount, or ask the recipient to send that amount, to TOPS. We also gladly accept donations. We know life is busy, but please be sure to follow through on your good intentions promptly. It will only take a few minutes, and you'll know you did the right thing!

Schools and Districts:

You may wish to use this curriculum in several classrooms, in one or more schools. Please observe the following:

PURCHASE option: Order this book in quantities equal to the number of target classrooms. If you order 5 books, for example, then you have unrestricted use of this curriculum in any 5 classrooms per year for the life of your institution. You may order at these quantity discounts:

02-09 copies: 90% of current catalog price + shipping.

10+ copies: 80% of current catalog price + shipping.

ROYALTY option: Purchase 1 book *plus* photocopy or printing rights in quantities equal to the number of designated classrooms. If you pay for 5 Class Licenses, for example, then you have purchased reproduction rights for any 5 classrooms per year for the life of your institution.

02-09 Class Licenses: 70% of current book price per classroom.

10+ Class Licenses: 60% of current book price per classroom.

Workshops and Training Programs:

We are grateful to all of you who spread the word about TOPS. Please limit duplication to only those Job Cards you will be demonstrating, and collect all copies afterwards. No take-home copies, please. Ask us for a free shipment of as many current **TOPS Ideas** catalogs as you need to support your efforts. Every catalog contains numerous free sample teaching ideas.

ISBN 0-941008-51-7

CONTENTS

 **PART I** **Preparation** *and* **Support**

4 Doing Stuff: *A Developmental Approach*
5 Testing the Waters: *Beginning with Confidence*
6 A Glossary of Basic Materials and Procedures
11 Dear Family Members…
12 How to Organize Work Space and Materials
13 Seven Easy Rules
14 Structuring Free and Independent Learning
15 Borrowing Card
16 Fill Your Class with Famous Scientists

 **PART II** **Reproducible Student Materials**

18 A/ POUR
35 B / SEARCH
48 C / COMPARE
68 D / DESIGN
84 E / MEASURE
100 F / DIVIDE
119 G / CALIBRATE

Doing Stuff: *A Developmental Approach*

Watch children interact with their world, and you will see them grasp and poke and fiddle and manipulate. This happens nonstop, or so it seems, unless they are sleeping. What's going on?

You are witnessing, first hand, the serious and necessary work of developing human brains, doing what brains need to do in order to survive. These children are literally learning how to think, building and internalizing mental structures to facilitate abstract thought. Without first moving physical objects with their hands, children will never learn to move mental images in their heads. Elementary education in a nutshell is simply this: Kids need to *do stuff!*

" *...self-organizing, autonomous learners need to do their own stuff to mature into well developed physical, emotional and intellectual human beings.* **"**

Children are obviously doing stuff in today's schools. But is it the *right stuff*? Why do first graders, curious and ready for anything, too often turn into passive-aggressive seventh graders who fear math, hate science and misunderstand technology? In an effort to efficiently cover course requirements and meet state standards, well-intentioned educators have created a top-down, adult-centered approach to teaching that forces classrooms full of children to do the *same stuff* in the same way, as prescribed by adult lesson plans. This teacher-knows-best methodology may cover the syllabus, but it is not what students (especially younger students) want or need.

Education at its best happens on a knife edge between structure and freedom. Teachers know best when it comes to structuring and maintaining a safe, nurturing educational environment. But students also know best when it comes to making their own personal learning decisions. May we, as teachers, humbly remember that self-organizing, autonomous learners need to do their *own stuff* to mature into well-developed physical, emotional and intellectual human beings. Indeed, the most admirably self-actualized students become so because they are encouraged (and taught) to follow their own interests.

Teachers know how important it is to honor individual differences, to care for children at both ends of the ability spectrum as well as all those in the middle. But with so many children to teach, and so little time to do it, we sometimes falter. When class control begins to slip, we tend to revert to a safer top-down style of teaching that requires everybody to do the *teacher's stuff.*

This book invites you and your students to do a special kind of *TOPS stuff* with lentils. The Job Box provides a certain kind of *unstructured structure*. Its four walls define unambiguous limits. Yet, inside, the lentils remain fluid and malleable. An astonishing range and depth of concrete experience is possible in the pouring, searching, comparing, designing, measuring, dividing and calibrating of stuff.

Put a box of lentils in front of a child, and she will show you what she needs to learn. Throw in a few thoughtfully constructed manipulatives, and she will spontaneously develop her own course of academic study, from simple basics to sophisticated complexity! The brain knows *how* to develop. This wisdom is deep and intrinsic. Trust it. You can assist, but never control. This means that you teach to the child, and not the lesson. Let's see how this works:

▶ *Chris, which Job Card would you like to try today?*

▶ *OK. Go ahead and set up your Job Box. I'll return as soon as I can to see how you're doing.*

▶ *I see you're working with containers of different sizes. Tell me what you have been doing.*

▶ *May I pour the lentils you put in this container into that one over there? No? Well, how can I help?*

▶ *OK. I'll hold the funnel while you pour. Well look at that! Which container holds more?*

▶ *Do you think this short fat container holds more than this tall skinny one? How can we find out?...*

In typical interactions like these, the lentil box provides the structure. The child pursues his own interests within prescribed limits. And the teacher, through simple, direct questioning, exploits learning opportunities that pop up moment by moment. The child's mind is focused, ready to capture new insights, build new mental structures, grow smarter. Mental fusion over a box of lentils. A transfer of insight and knowledge, the *good stuff* of good education.

Testing the Waters: *Beginning with Confidence*

You are visiting Lentil Lake for the first time, and wonder whether you might go for a swim. First you dip your fingers and toes. If the water feels warm enough, you decide to wade in up to your waist. If the lake bottom feels firm and friendly, you take courage and decide to plunge in all the way. Next year, when you revisit these waters, you might decide to take a running dive. But for now, you want to test the waters of Lentil Lake step by step.

• *Fingers and Toes:*

Set up a single learning station, with one Job Box, in a corner of your room. Introduce the first Job Card to your entire class, using the materials indicated on the card. The first chapter is a nice place to begin, but you could start with another chapter just as well.

A single "person" icon on some Job Card headings indicates that the activity is best done individually. But two students *might* successfully work together, if they are good at sharing.

Students use this learning station, by permission, when their other work is complete. With many children and only one Job Box, working with lentils will be seen as a privileged activity, and students will be especially motivated to behave well. This might be a good time to introduce some of the Seven Easy Rules on page 13.

• *Waist Deep:*

When you are comfortable with the water temperature in Lentil Lake, **set up multiple learning stations**, with multiple Job Boxes. Each day or so, you might wade in a little deeper by adding another Job Box and/or a new Job Card. As you run out of dedicated space for learning stations, allow model students to retrieve materials from a designated storage area, then work at their own desks, on the floor, or wherever practical.

It is not necessary to do these chapters in order, or finish one chapter before moving on to the second. Quite the opposite. We strongly recommend that you introduce lessons horizontally, offering the first Job Card in every chapter first before introducing the second. This maximizes variety, providing students with rich developmental choice, and reduces the need to duplicate materials.

• *Take the Plunge:*

Are students becoming familiar with the rules? Do they handle equipment properly, clean up lentil spills and put everything back where it belongs? Are our recommended class procedures falling into place as you introduce them? Do students feel confident? Do you feel confident? Have you learned to trust the integrity and design of our program? Then you are ready to **mainstream the curriculum** into your class as a regularly scheduled science period.

No longer is our program supplementary, something to do at a learning station when other work is complete. Lentils have worked their way into your confidence (and into the corners of your classroom) as an important medium for learning. Now everybody does science together, in a thoughtfully structured program that honors diversity and maximizes free choice. Job boxes are scattered all around your classroom, with students working individually or in pairs.

And, even though this is *science* period, the *activities are so interdisciplinary*, your students are also learning math, language arts, social studies and art. Continue to work horizontally and vertically through the book. Feel free to suggest variations that occur to you, or substitute suitable materials. Improvisation is at the heart of creativity!

Within a few months, your students will be ranging far and wide through the curriculum, happily engaged in designing their own projects, initiating deeper investigations into their areas of greatest interest, moving beyond the Job Box. Ah, yes. The water in Lentil Lake is really fine.

GLOSSARY *of Basic Materials and Procedures*

All supplies used in more than one chapter, and instructions for turning them into manipulatives for student use, are listed below. All tools needed to make any item in this book are also listed here. Items dedicated for use in single chapters can be found on special materials pages in those chapters. Any time you find an item underlined, here or elsewhere, look for a more detailed explanation in this glossary.

Some items are needed in quantity. We begin these descriptions with two numbers separated by a dash and ending with a colon, *i.e.* 2–4: . The first number estimates what you'll need for a single learning station set up in the corner of your classroom or at home. The second estimates what you'll need to accommodate 30 students working alone or in pairs during a dedicated science period.

baby food jars (BFJ's)

2–4 sets of 3 baby food jars: A trio of small (2½ oz), medium (4 oz) and large (6 oz) baby food jars. Gerber brand calls these 1st, 2nd and 3rd foods, respectively. Masking tape labels: *small, medium, large.*

blank paper

Used for student drawings. Recycle the back sides of unused photocopies. Like pencils and paper, this item is understood to support Job Cards even though it is not specifically itemized as a needed material.

booklets

1. Photocopy the associated line master. Fold it along the grey center line.

2. Trim off the long border, parallel to the fold, cutting through both layers of paper.

3. Cut *toward* the fold (to prevent fanning) on the dashed lines between the 8 folded pages. Discard the trim.

4. Order the pages, folded ends to the right, with page 1 on top. Page numbers will be in the upper right corner.

5. Crease the folded edges of the collated pages. Jog them even, then bind the uneven cut edges on the left with a single staple.

bottle caps

37–37: Use twist-off caps that are not bent. If beer labels are offensive, cover with masking tape.

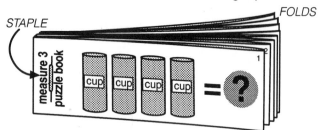

bottle lids

5–50: These are especially important for sealing lentils in liter storage bottles away from moths, mice and other pests. They also close half-liter bottles (labeled "small bottle" in this curriculum). Some threads may not match perfectly, but still work well enough. All bottle lids must measure 1⅛ inch (3 cm) in diameter. Some

manufactures make bottles with wider mouths. Avoid these completely. When not sealing containers, store lids in a container of appropriate size and label: *bottle lids.*

box and brick

A thrifty filing cabinet. Use to store student folders.

cleaning screen

Use ⅛ inch mesh hardware cloth (wire screen). Frame a 6 inch square (or somewhat larger piece) in electrical tape to cover wire ends. Bend up the perimeter on 3 sides to form a shallow dust-pan shape. Use this to clean lentils swept off the floor. Shake the screen over a waste basket to filter out dust and dirt. Pick out larger refuse by hand. If swept-up spills are dumped directly into the Job Box, your lentil supply will grow dirty and trashy over time. (*NOTE:* When purchasing this material, get some ¼ inch mesh as well. See *screen*, page 36.)

cleanup

Use any procedure that works for you. Here are some options. ✓Lentils are big enough to pick up off the floor in moderate numbers. These are still clean enough to return straight to the Job Box. ✓Sweep larger spills into a cleaning screen, filter and return to the Job Box. ✓Provide a wide-mouth holding container labeled "dirty lentils." Direct students to return all swept-up lentils here for later screening by a student helper. ✓Discard swept or vacuumed lentils, replacing as needed. Lentils are, after all, quite inexpensive and totally biodegradable.

clear cups

12–20: These are tall, tapered, clear plastic, 10 ounce cups used both in lentil activities and as storage containers. Solo and Polar brands are suitable.

clear tape

Use ½ inch or ¾ inch wide rolls. This is used only in a few Job Cards and is generally dispensed by the teacher.

craft sticks

60–70: Also called popsicle sticks. Available by the box at craft stores. Store in a clear cup.

crayons and colored pencils

These are optional student materials. Drawing and shading tasks are usually done with graphite pencil.

cups

When we refer to a cup (or fractional cup), we mean amber pill vials (often cut to a particular size). See also <u>standard cups</u>, <u>cut-to-size containers</u>, and <u>plastic vials</u>.

cut-to-size containers

All volume measure in **Lentil Science** is based on 60-dram <u>plastic vials</u>, commonly used by drug stores to package prescription drugs, and also available from TOPS. Even though 60 drams is officially equivalent to only 94% of a real cup, this is close enough to look and feel like a real cup. More important, all our related measures (pints, quarts, half cups, etc.) derive from this standard.

And how will you derive them? Some containers can be used as manufactured. Other need to be cut to size, using templates we provide. And others can be sized experimentally by pouring lentils.

Suppose, for example, you need to create a pint that holds precisely 2 <u>standard cups</u>. Follow the instructions below. (Then generalize this procedure to derive other standards you need to make.)

1. Work on a level surface. Overfill a <u>standard cup</u> with lentils. Shake once to shed excess lentils off the top. You now have a slightly rounded, loosely packed cup of lentils that we call "fair and full." <u>Funnel</u> this cup into a <u>half-liter bottle</u>. Add a second cup in the same manner. Tilt the bottle, if necessary, to level the contents, but don't shake it down.

2. Mark the level of the lentils with a dot from a permanent marker on the outer surface. (Note: If your container is translucent, hold it at eye level against strong light to see the lentils inside. If the container is opaque, bring your fingers together inside and outside, level with the surface of the lentils. (It's OK if this reference dot is only approximate.)

3. Hold the marker absolutely steady against your reference dot: brace your hand on a stationary object of suitable height. A pile of books works well.

4. Rotate the container against the steady marker tip with your free hand, inscribing a level line around the container, back to your starting dot. (Note: If the container is square, a milk carton perhaps, slide it past the marker on all four sides.)

5. Inscribe two more lines around the container, one a little above your reference dot, and another a little below. (These extra lines are not necessary if your original line is certain.)

6. Cut around the top line with curved <u>toenail scissors</u>. Test your container at this height by filling a standard cup fair and full twice, and pouring into the pint you are testing. Cautiously trim parallel to lower lines as necessary, until the level is precise. (Note: If the container is small and curved, you may wish to spiral the cut downward from the top, converging with your inscribed line. Always hold the scissors so the blades curve *away* from the line you are approaching.

7. Always feel for sharp edges and snags. If the edge remains too sharp, cover with masking tape. (We have never found sharpness to be a problem.)

electrical tape

Use a roll of 3/4 inch black plastic tape. This teacher supply is used in small quantities to prepare student materials.

equation tags

Photocopy the line master on page 93 and assemble as directed in the grey section. Store these tags in a clear cup. Masking tape label: *equation tags*.

extra cups jug

1–1: Whenever a Job Sheet specifies "extra" cups (or an "extra" half cup), students should retrieve plastic vials from this jug, leaving *dedicated* cup sets related to other chapters (Compare, Measure and Divide) intact for other Job Cards.

Photocopy "extra cups" label(s) on page 85.

Prepare a <u>gallon storage jug</u> to hold 7 <u>plastic vials</u>. Six of these should be 60-dram vials labeled "cup" with masking tape. The seventh should be a 30-dram vial (or taller vial cut to size) labeled "half cup." Fix a paper clip to the side of each whole cup as follows:

1. Roll masking tape into a small tube, sticky side out. Center it, with open ends aimed up and down, on the side of the cup.

2. Stick a new (not bent) paper clip on the tape as illustrated. The ends should point down, with the double loop projecting just above the edge of the tape.

3. Tape tightly over the top, matching the top edge of the rolled tape underneath.

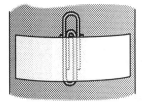

film cans

6–6: Use plastic film canisters with snap-on lids.

floor cleaning equipment

Use any of the following: a broom, whisk broom, dust pan, <u>cleaning screen</u>, carpet sweeper or vacuum.

four nesting containers

<u>Cut to size</u> clear plastic beverage bottles of different diameter, but equal height. These sizes nest well:

Narrow: Use a <u>half-liter bottle</u>. Since this is the shortest bottle, cut it to maximum height, just below its sloping shoulders. Then cut the other 3 bottles to this same height.

Masking tape label: *narrow.*

Medium: Use a 24 ounce water bottle (found in grocery or convenience store cold cases). Best shape is tall, with straight sides.

Masking tape label: *medium.*

Wide: Use a <u>liter bottle</u>, perhaps a left-over bottom reserved from cutting a funnel.

Masking tape label: *wide.*

Very Wide: Use a 2-liter soft drink bottle. (These may also be recycled as <u>funnels</u>.) Use 2 masking tape labels, one above the other.

Top label: *4 nesting containers.* Lower label: *very wide.*

funnels

1–20: Cut these from one-liter or two-liter bottles. Because you'll need as many as 40 matching <u>liter bottles</u> for lentil storage, have them in hand before cutting up extras for funnels. Cut mismatched or odd-shaped bottles first. Or make funnels from 2-liter bottles, which are used for little else in this curriculum.

Directions for cutting bottles of either size are the same: Measure down the sloping shoulder about 2 inches from the base of the neck, and make a dot. Inscribe a circle at that level as in <u>cut-to-size</u>. Weight the bottle with lentils or water, if needed, to help steady it as you turn it. Pierce the bottle with pointed toenail scissors and cut along the inscribed line. Your cut will be somewhat ragged; trim as necessary to make it even. Feel for sharp snags. Funnels from either bottle stand equally tall, measure about 9 cm in diameter, and funnel nearly equal capacities.

gallon storage jugs

5-11: Cut the widest possible opening in the top of each jug with toenail scissors. Keep the handle intact.

hacksaw

Needed for material preparation only in A/ Pour.

half-liter bottles

5-7: These are found in grocery stores and convenience store cold cases. The best shape is tall and straight. Ribbed sides are OK. Bottles should look identical. If the screw-on lids also fit the 1-liter bottles, this is ideal.

Masking tape labels: *small bottle.* (Don't label it "half liter.")

index cards

14-14: Use 4 x 6 inch cards.

job boxes

1-20: We used 19 x 14.5 x 3 inch (49 x 37 x 8 cm) corrugated cardboard boxes, commonly supplied by nurseries to carry home potted plants. These are cut and scored to fold into self-locking boxes, with seamless bottoms that are especially good for scooping and pouring lentils. You might negotiate a special price for a bundle of unfolded boxes, then assemble them with your students as a class project.

Cover all inside seams – anywhere lentils might stick or hide – with clear packaging tape. Keep the bottom entirely free of tape. (It tends to peel up over time, as students scoop up lentils.)

Plastic tubs or sorting trays with similar dimensions also work.

Add a holder for the Job Cards: Cut a strip of packaging tape about 3/4 the length of one of the longer sides. Center it inside the box on this longer side, flush with the bottom edge. Cut a 6 inch (15 cm) length of drinking straw. Fix it horizontally on the taped side, about a finger width from the bottom of the box, with masking tape at each end. Test to see that a folded Job Card slips easily into this holder.

lab coats and ID badges

Recycle white shirts and photocopy ID badges as outlined on page 16. These props are optional for establishing alter-ego famous-scientist identities.

lentils

4-80 pounds: Say what? No, this is not a typo. About 20 Job Boxes, on average, are used by a class of 30 students working alone or in pairs. Each Job Box typically requires 1 or 2 liters of lentils to support Job Card activities, and each liter bottle holds nearly 2 pounds.

I can see you now, dear reader, closing your eyes and imagining how scary it might feel to be stuck in a classroom with 30 youngsters and 80 pounds of lentils. Please see *Testing the Waters* on page 5 to find our comforting suggestions for introducing **Lentil Science** at *your own* pace.

Eighty pounds of lentils seems strange only because modern classrooms are generally bereft of volumetric experiences. Except for occasional rice tables on the K-1 level and demonstrations with water at higher levels, children have very few opportunities to experience the qualitative and quantitative character of 3-dimensional space. **Lentil Science** addresses this problem in a thoughtful, organized way. You *can* manage it. And your kids will be *way* smarter for it!

Find lentils, sold in 1 or 2 pound bags, near the rice and beans at your local grocery store. Or purchase them in bulk at wholesale food outlets. Or strike a special high-volume discount deal with your grocer. Or ask for help from family members (see page 11 for a letter you can send).

Always store lentils in closed bottles when not in use, away from moths, mice, and other pests.

liter bottles

5-50: Collect both plastic bottles and bottle lids. Most will be used for lentil storage, some for Job Card experiments. These commonly package ginger ale, seltzer water, and other carbonated drinks appealing to adult tastes. Try to collect a matching set of bottles with equal height and width. Simple, smooth-sided cylindrical shapes work best. Reject ribbed or unusually-shaped bottles unless nothing else is available. Odd liter bottles can be cut into funnels.

If you can't locate enough one-liter bottles right away, substitute the more ubiquitous two-liter soft drink bottle as temporary holding containers. These larger bottles are probably too flimsy to hold up over the long term, and are awkward to use.

magnets

4-16: Rectangular ceramic refrigerator magnets used in B/Search. See page 36.

manila file folders

8-37: Use folders sized to hold $8\frac{1}{2}$ x 11 inch paper. These organize curriculum materials and student work.

masking tape

Use standard $\frac{3}{4}$ inch wide rolls. This is used only in a few Job Cards and is generally dispensed by the teacher.

medium-sized cans

3-3: Use clean 15 or 16 ounce vegetable cans, labels removed, bottoms intact.

measuring bottles

1-3: Select these from your supply of liter bottle(s). Prepare as follows:

1. Run a strip of masking tape down the bottle from "neck" to "foot."

2. Fill a standard cup *fair and full* with lentils; pour it into the bottle; tilt if necessary to level the contents; mark the level of the lentils with a *pencil* line on the tape.

3. Calibrate in this manner up to 4 cups, maintaining a loose pack. (Do *not* settle the lentils by "tickling" or tapping the bottle. If settling does occur, invert the bottle once to reestablish a loose fill, and tilt to level the contents inside.)

4. Label the center of each pencil line with a black permanent marker: 1C, 2C, 3C, 4C. Write "loose pack" at the top to remind students that these lines represent "fluffy" lentil volumes.

5. Ring the neck with a dedicated rubber band. (It will be used to predict lentil heights in some Job Cards.)

packaging tape

Use clear, 2-inch-wide rolls. Opaque tape is *not* a good alternative. This is an important teacher supply used widely to prepare student materials.

paper clips

Use one box of standard sized clips.

paper plates

1-4: Use a generic, 9-inch diameter picnic plate.

paper punch

Teacher use only.

permanent marker

Teacher use only. Select a fine point with black ink. "Sharpie" brand works well.

plastic vials

We used prescription pill vials manufactured by Owens-Indiana, with an "O-I" stamped on the bottom, as well as the letter "T" followed by a number indicating its volume in drams. The tops are rimmed by "child-proofing" nubs. We selected these because of their wide availability in national drug store chains. You can also purchase them directly from TOPS in all needed sizes.

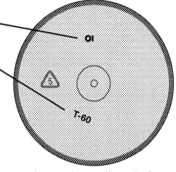

The first quantity given for each item (for a single learning station) includes enough vials to produce one complete set of materials. These include milk-jug container sets for Compare, Measure and Divide, an extra cups jug set, plus a few additional vials (each dedicated to a special use).

The second quantity given (for a class of 30 students) includes one set of everything above, plus two additional sets of vials (three sets in all) for Compare, Measure and Divide:

12-22:	60 dram
1-3:	40 dram
4-10:	30 dram
1-3:	20 dram
3-9:	16 dram
2-6:	13 dram
1-3:	8½ dram

rubber bands

30-35: medium-sized rubber bands. Store in a clear cup.

scissors

A pair of adult scissors for teacher use, and safety scissors for student use.

scoops

1-20: Use quart or pint cardboard milk cartons. (Half pint school-lunch cartons make an acceptable, though somewhat smaller scoop.) Cut to size: cut off the top with toenail scissors, leaving a box about 1½ inch (4 cm) tall. Cut away one of the sides. (Alternatively, you might leave all four sides intact. A box scoops more efficiently and pours well, but doesn't have the classic shape.)

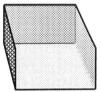

This important tool is always needed for cleanup, to return lentils to storage bottles. It is only specified on a Job Card, however, when used as an integral part of the investigation.

standard cup

12-22: This term refers to 60 dram Owens-Indiana plastic vials rimmed with lock-tight nubs. Their volume is the standard by which all other measuring containers in this book are calibrated.

stapler

One on the teacher's desk is enough.

story paper

This has a large frame for artwork at the top, and lines for writing underneath. It can be used in many activities as a writing and drawing supply, but will only be listed on those Job Cards where it assumes special importance. Find the line master on page 40, associated with Job Card B/2 for first use. Or you can substitute ready-made tablets of similar design.

straws

1-20: plastic drinking straws. Used to hold Job Cards in Job Boxes.

toenail scissors

These must be curved and pointed. They are used by the teacher for cutting plastic and cardboard containers to size. Heavy duty scissors work best.

tubs

4-8: Try to locate a uniform set of 1 pound margarine or butter tubs. These generally hold 3 standard cups, as manufactured. This size is ideal; a slightly larger size is OK. Do *not* substitute smaller sizes.

white glue

Used by the teacher to prepare student materials. Apply from bottle or stick. Student use is generally optional.

wire cutters or shears

Needed to cut hardware cloth to size in B/Search.

wood saw

Needed to cut two 2x4's to size in B/Search.

writing paper

This item is understood to support Job Cards even though it is not specifically itemized as a needed material. Have your students use lined paper appropriate to their age group.

Dear Family Members of _____,
    We are setting up a hands-on learning program called **Lentil Science** *at our school. We will be using lots of lentils, bottles, and pill vials, to pour, search, compare, design, measure, divide and calibrate. Any help you might give us in purchasing lentils or gathering the recycled materials listed below is much appreciated. We need to collect most of these materials by* _____.

Sincerely,

QUANTITY	
Program requires:	**We currently need:**
20	
80	
70	
20	
11	
30	
22	
3	
10	
3	
9	
6	
3	

corrugated cardboard boxes with seamless bottoms, measuring approximately 19 x 14.5 x 3 inches. (49 x 37 x 8 cm). These are commonly supplied by nurseries to carry home potted plants.

pounds of lentils in bags of 1 pound or more.

clean, dry, one-liter plastic bottles, with **screw-on lids** that measure $1\frac{1}{8}$ inch (3 cm) in diameter. These commonly package ginger ale, seltzer water, and other carbonated drinks appealing to adult tastes. Simple, smooth-sided cylindrical shapes work best. Please don't confuse this 1.0 liter size with larger 1.5 liter and 2.0 liter sizes also on the market.

one-quart milk cartons made from cardboard, not plastic.

one-gallon milk jugs, plastic.

old white dress shirts to recycle as lab coats.

plastic pill vials, amber color with child-proof "nubs" around the rim. We don't need the lids. The brand we need is manufactured by Owens-Indiana. These have an "O-I" on the bottom, along with the letter "T" followed by a number showing its volume in drams. We are looking for the following sizes, listed from largest to smallest:

T-60 pill vials
T-40 pill vials
T-30 pill vials
T-20 pill vials
T-16 pill vials
T-13 pill vials
T-8$\frac{1}{2}$ pill vials.

BOTTOM OF PILL VIAL

O-I

T-60

Support: 1 copy per student

1. Prepare 56 Job Cards: Make one photocopy of each card. Shift the book on the copy window until you find the best position for a reasonably well-centered image. Fold each photocopy twice, first bisecting the *horizontal* gray line, then the *vertical* grey line. This creates a "greeting card" format, with student directions on the front of each card, needed materials on the back, and teaching notes handy on the inside. (Card corners and edges may not match as a result of imperfect centering. You may trim these edges even, if you wish, but no one is likely to notice.)

2. Prepare 7 Chapter Folders: Center the long side of a 4 x 6 inch index card along the spine of a standard-sized manila folder. Tape it with wide packaging tape around this spine, and up to the top of each short side. Leave the last long side open to form a pocket. Slip a chapter set of Job Cards into the pocket, and Job Sheets into the folder. Prepare 6 more Chapter Folders in the same manner, and label the tabs.

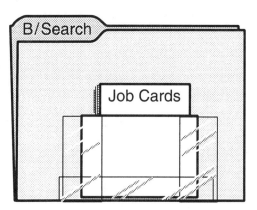

3. Photocopy Job Sheets: These pages follow some Job Cards to support written work. The label *Job Sheet* appears in the lower right corner of the page, followed by suggested quantities to photocopy (usually 1 per student). Paper clip duplicated sets together, and store in numerical order inside corresponding Chapter Folders.

Note: Photocopy only a few of each Job Sheet to start, until you develop a sense of how fast they are being consumed. You can always copy more later.

4. Prepare Student Folders: Assign each student a class number corresponding to the position of each name on your finalized class list.

Boldly label the tab on a standard-sized manila folder with each student's assigned number to facilitate quick and easy filing.

Photocopy class sets of pages 13 and 15. Staple the *Seven Easy Rules* and *Progress Chart* inside the front cover. Fold the *Borrowing Cards* in quarters, then paper clip one inside each folder.

At a later time you might have students write their names (real, or assumed *Famous Scientist* names) on the covers of their folders, and personalize them with decorative science themes.

5. Prepare Special Materials: Find detailed instructions at the beginning of each chapter for gathering and assembling these items. Photocopy each page of associated labels to tag special items. Use brightly colored copy paper if available. Store all chapter materials together near each chapter sign. (These require 1 square foot of space, or less, per chapter.)

6. Prepare Basic Materials: Send the *Dear Family Member* letter home with each child (see previous page) to enlist outside help in gathering *high-quantity basics*. Store these together, on and under a table or counter:

> Job Boxes
> liters of lentils
> bottle lids (store in a labeled container)
> scoops
> funnels

Store *low-quantity basics* together, near a stand-up sign or in a labeled box that reads "BASICS." Photocopy this sign on page 49 and gather the indicated materials. Apply the label(s) on page 85.

7. Organize Cleaning Materials: You might provide *some* of these in a corner of your room, depending on your <u>cleanup</u> strategy:

> <u>cleaning screens</u>
> <u>floor cleaning equipment</u>: broom, whisk broom, dust pan, carpet sweeper or vacuum
> a "dirty lentils" container

8. Issue <u>Lab Coats and ID Badges.</u> These are optional, for classes using alter-ego identities of *Famous Scientists* (see page 16). Medium brown paper bags, labeled by class number, may help you organize coats and badges.

Seven Easy Rules

1. Be safe.
• Never throw lentils, or anything else.
• Never put lentils in your mouth, ears or nose. (Lentils in ears may need a doctor to remove.)

2. Respect others.
• Imagine you are in a bubble as big as your arms reach. This is your personal space. Don't enter another person's bubble unless they invite you.
• No put-downs or name-calling.

3. Avoid spills. Clean them up.
• Work inside your Job Box.
• Clean up all spills as soon as they happen.
• You may tip a Job Box to gather lentils, but it should always touch your table. No Job Box may be carried around unless it is empty.

4. Stay on task.
• Display the Job Card you are working on.
• Use only the materials pictured on the Job Card. (You will often need to return empty liter bottles and lids to storage. Sometimes the scoops and funnels should stay in storage, too. Get them when you are ready to clean up.)

• If you want to do your own experiment, display Job Cards that say "On your own."

5. Finish what you start.
• Finish old work, and ask for a checkpoint, before starting a new Job Card.
• You have "first dibs" when continuing a Job Card the next day, as long as you don't keep it too long.

6. Take care of equipment.
• Put everything back where it belongs, so everyone can find what they need next time.
• Keep lentil storage bottles closed.
• Don't keep equipment for yourself. It belongs to all of us. When it's gone, everyone loses.

7. Eat a "balanced diet."
• Work on many different Job Cards, both those you love to do best **and** those you need to learn.
• Each time your teacher OK's a finished Job Card, fill in *one* white line in the correct box on the Progress Chart below. Each box holds up to 4 marks. Can you earn a mark in every column? In every box? That's good balance!

Progress Chart *for* _____

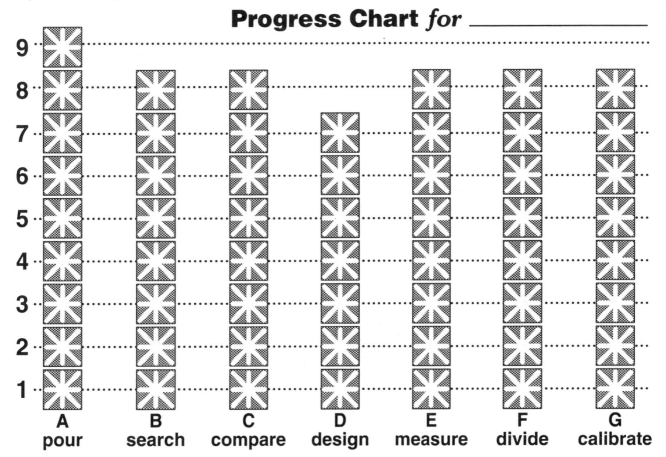

Model these steps for your students, then ask them to role play until you're sure they've got it. Review as necessary.

Every science period:

1. Look inside your Student Folder. You'll find Seven Easy Rules, a Progress Chart, and a Borrowing Card with a calendar on it. This is where you'll put all your written work.

2. Take your Borrowing Card to the Chapter Folders, and choose a Job Card to work on. (If you haven't finished work on your last Job Card, or didn't get a teacher check, go back to that one.)

Note: Allow students who wish to continue with an unfinished Job Card to choose first, as long as they're not monopolizing a popular card. All remaining cards then become "free" for others to borrow. (Only those cards that you have already introduced should be in circulation. Keep adding new ones day by day.)

3. Borrow a Job Card:

• Write its *Letter/Number* in today's box on your Borrowing Card.

• Put your Borrowing Card in the pocket of the Chapter Folder.

Note: Borrowing allows you to see at a glance who is using which card. And Job Cards are less likely to get lost if students must return them to get their Borrowing Cards back.

4. Find out how many people should work on that Job Card. Little "person" symbols on the front of each folded card show if you should work alone or with a partner. If you find someone to work with, put their Borrowing Card in the Chapter Folder pocket along with yours.

5. Gather materials. Bring everything you see on the back of the Job Card to your work area. Each item has its own special place. If Job Sheets are shown, find these in the Chapter Folder. Look for all materials near the chapter sign (or in the chapter box) and near the "basic" sign (or in the "basic" box). Finally, get a Job Box and however many bottles of lentils your Job Card shows.

6. Stand the Job Card in the holder at the back of the Job Box. Pour in lentils if the Job Card tells you to. Go to work. Experiment. Ask for a teacher check any time you're ready.

7. Ask for a checkpoint when you finish your work. When your completed work has been OK'd, mark a line on your Progress Chart. (Colored pencil looks nice.) If you have time to start a new Job Card, trade the old one back for your Borrowing Card and find a new one to record and swap. (Squeeze the extra number into today's box, or use tomorrow's box.) Return old materials and gather new materials as required.

Note: Initial your approval at the top of each student's work. This might be a completed Job Sheet, a story, answered questions, a drawing, a computation, just about anything. Even very young children can produce some kind of recorded response. At the very least, they can copy the number of the card they have been working on for your initialed approval. Gently nudge each individual toward a higher personal standard, greater elaboration, more refinement as they grow and develop.

8. Follow the rules stapled in your Student Folder. Be nice. Practice care and respect for each other and for the equipment.

Note: You can reduce theft in your classroom by instilling a sense of group ownership. Introduce equipment with care and reverence. Emphasize how you made each new item or where you purchased it, how it works, and where it belongs. Then explicitly gift it to all class members: I give this to you, and you, and you..., to help you learn and grow.

A "balanced diet" is also very important. Respect the deep wisdom of each child's mind and body to seek nourishment in its own natural way. But watch out for those children who have already become "sugar junkies," looking for fun, taking the easy way out. Use your adult perspective and wisdom (and stickers and stars!) to gently coax these children to work broadly through all subject areas, earning many marks across their Progress Charts. As they do this, you'll enjoy the added benefit of fewer materials serving more students. Everybody wins!

9. Please clean up when I tell you to. Be careful, courteous, quiet, and quick. Put everything back where it belongs. Return the Job Card to the correct Chapter Folder pocket, and take back your Borrowing Card. Put *all* of your papers, including Job Sheets, artwork, stories and such, in your folder.

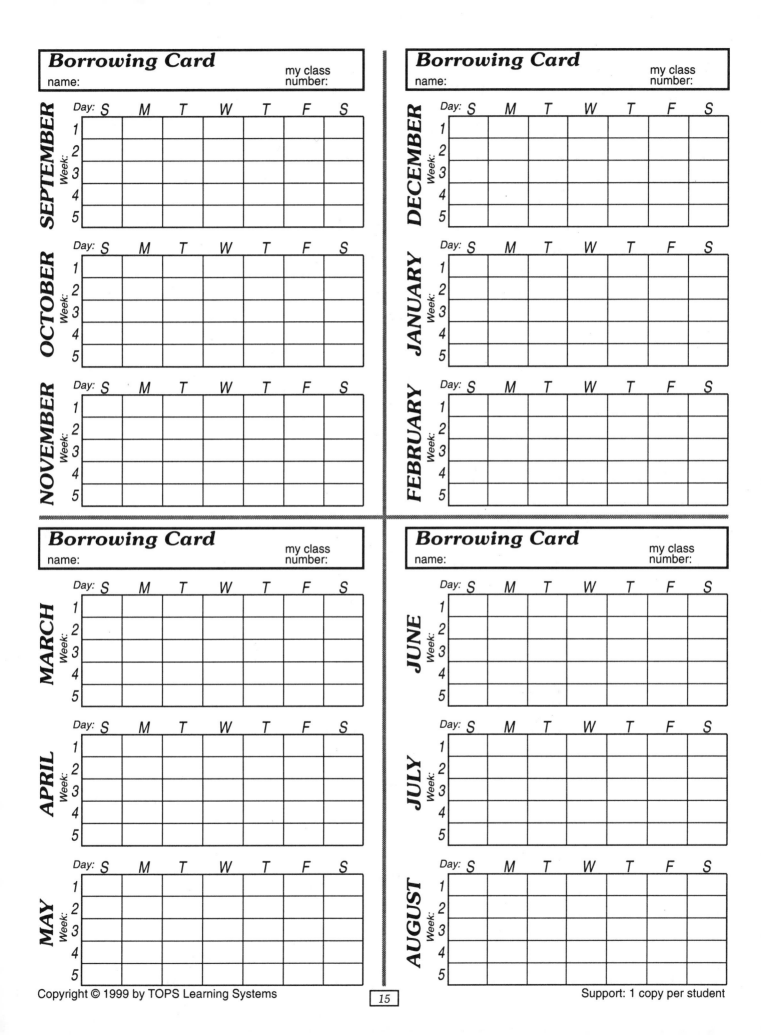

Fill Your Class with Famous Scientists

Ego often gets in the way of learning. We all know this. We see it in our students and feel it in ourselves. We *want* to learn new things, of course. But first we have to protect our self-image – make certain that our ignorance, real or imagined, is never discovered.

Lozanov, a Bulgarian psychologist and language teacher, called this problem a *learning barrier*. He found that he could sidestep problems of ego by allowing his language students to role-play new identities. You can, too. Try this in your class:

1. Photocopy the names of famous scientists on the next page. Read some of the famous names from this list to your class. Post a copy on your bulletin board or hand out duplicates. Ask students to think about who they might like to be.

2. Ask each student to choose a name from this list. Girls can cross the gender gap by adopting a feminine version of any male name. Two or more students might choose the same name: *Issac Newton I* and *Isabel Newton II*, for example. Some may wish to take the names of more recent scientists not on the list: Carl (Carla) Sagan, Linus (Lucy) Pauling and many others.

3. Photocopy, cut out and distribute the ID badges. Students should fold them in half, neatly fill in both sides, then sign the bottom. Laminate these tags (or cover front and back with clear packaging tape), paper punch, and provide heavy string or yarn for wearing around the neck.

4. Ask each student to introduce his or her scientist-self to the class and talk briefly about background and accomplishments. You might assign students to research their "autobiographies" in the library or on the internet. At the end of each presentation, students might hold up their signed card as evidence that they have "taken the pledge."

5. Issue lab coats (old white shirts) for students to wear with their name tags.

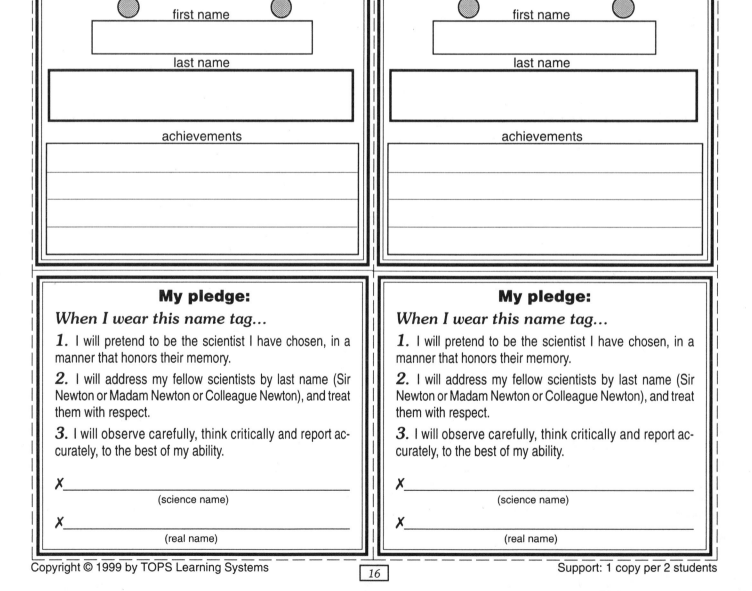

first name

last name

achievements

My pledge:

When I wear this name tag...

1. I will pretend to be the scientist I have chosen, in a manner that honors their memory.

2. I will address my fellow scientists by last name (Sir Newton or Madam Newton or Colleague Newton), and treat them with respect.

3. I will observe carefully, think critically and report accurately, to the best of my ability.

X_____
(science name)

X_____
(real name)

---

first name

last name

achievements

My pledge:

When I wear this name tag...

1. I will pretend to be the scientist I have chosen, in a manner that honors their memory.

2. I will address my fellow scientists by last name (Sir Newton or Madam Newton or Colleague Newton), and treat them with respect.

3. I will observe carefully, think critically and report accurately, to the best of my ability.

X_____
(science name)

X_____
(real name)

Support: 1 copy per 2 students

50 Famous Scientists

Howard (Holly) **ALKEN:** developed first large-scale digital computer

Amedeo (Amedea) **AVOGADRO:** determined Avogadro's number of atoms in a mole

Alexander (Alexandra) Graham **BELL:** patented the telephone

Daniel (Daniela) **BERNOULLI:** studied fluid movement and pressure; explained how airplanes fly.

Henry (Henrietta) **BESSEMER:** developed a process for steel production

Niels (Nielsa) **BOHR:** leading developer of the quantum theory

Robert (Roberta) **BUNSEN:** developed the Bunsen burner

George (Georgia) Washington **CARVER:** agricultural chemist

Henry (Henrietta) **CAVENDISH:** discovered hydrogen

Madam (Sir) **CURIE:** noted for work with radium

John (Joan) **DALTON:** developed the atomic theory

Charles (Charlotte) **DARWIN:** developed the theory of organic evolution

Thomas (Tamara) **DOOLEY:** jungle doctor

Rudolf (Rhoda) **DIESEL:** patented the Diesel engine

Christian (Christine) **DOPPLER:** demonstrated the Doppler effect

Thomas (Thomasina) **EDISON:** inventor of the electric light bulb and phonograph

Albert (Alberta) **EINSTEIN:** developed the theory of relativity

Leonhard (Leonora) **EULER:** authored the first calculus book

Gabriel (Gabreille) **FAHRENHEIT:** introduced the Fahrenheit temperature scale

Michael (Michelle) **FARADAY:** noted for work with electricity

Alexander (Alexandra) **FLEMING:** discovered penicillin

Sigmund (Sigrid) **FREUD:** founder of psychoanalysis

GALILEO: founder of the experimental method

Robert (Rhonda) **GODDARD:** founder of modern rocketry

William (Wilma) **HARVEY:** discovered circulation of the blood

Edwin (Edwina) **HUBBLE:** discovered evidence of the expanding universe

Julian (Julie) **HUXLEY:** noted philosopher of science

Edward (Edna) **JENNER:** vaccination pioneer

Carl (Carla) **JUNG:** pioneer in analytical psychology

Sister Elizabeth (Brother Ellis) **KENNY:** developed treatment of polio

Johannes (Johanna) **KEPLER:** developed laws of planetary motion

Antoine (Antoinette) **LAVOISIER:** founder of modern chemistry

Louis (Louise) **LEAKEY:** discovered fossil remains of early hominids

James (Jane) Clerk **MAXWELL:** developed Maxwell equations of electromagnetism

Maria (Marion) Goeppert **MAYER:** discovered structure of the atomic nucleus

Gregor (Greta) **MENDEL:** discovered heredity, dominant and recessive genes

Albert (Alberta) **MICHELSON:** established the speed of light as a constant

Robert (Roberta) **MILLIKAN:** investigated electronic charges and the photoelectric effect

Thomas (Tamara) Hunt **MORGAN:** developed the chromosome theory of heredity

Isaac (Isabel) **NEWTON:** discovered laws of gravitation and motion

J. Robert (Robin) **OPPENHEIMER:** developed the atomic bomb

Louis (Louisa) **PASTEUR:** developed the process of pasteurization

Max (Maxine) **PLANK:** originated the quantum theory

Joseph (Josephine) **PRIESTLEY:** discovered oxygen

Walter (Wanda) **REED:** proved mosquitoes transmit yellow fever

Bertrand (Bernadine) **RUSSELL:** philosopher, founder of modern logic

Ernest (Ernestine) **RUTHERFORD:** discovered the atomic nucleus

James (Jane) **WATT:** invented the steam engine

Norbert (Nora) **WIENER:** founder of the science of cybernetics

Ferdinand (Fern) **ZEPPELIN:** airship designer.

Support: several copies

A / POUR

In this chapter: Pour out lentils. Scoop them back. Turn an hourglass over and over. Reversible processes repeat to measure out time. Experience the sequence; order the sequence; draw the sequence (what you *really* see, not what you think you see). Listen. Look. Touch. Write creatively and metaphorically.

job card **1**	pour a mountain	21
preparation	movie frames	22
job card **2**	pour and stop halfway	23
job sheet	draw how lentils pour	24
job card **3**	build a tower	25
job card **4**	pour the same amounts	26
job sheet	draw how equal looks different	27
job card **5**	look, touch, listen	28
job sheet	how lentils look, feel, sound	29
job card **6**	divide equally	30
job card **7**	set up a still life	31
job sheet	draw lentils at rest	32
job card **8**	make an hourglass	33
job card **9**	on your own (no hands)	34

Basic Materials: Quantities define maximums needed to support any one Job Card in this chapter. Store *high-quantity basics* (Job Boxes, liters of lentils, bottle lids, scoops, funnels) on and under a table or counter. Store *low-quantity basics* near the "basics" sign (see page 49) or in a "basics" box. Consult our Glossary on pages 6-10 for a full description of these items. See the next page for additional special materials used in this chapter.

- [ ] **1 job box**
- [ ] **up to 2 liters of lentils**
- [ ] **1 scoop**
- [ ] **1 funnel**
- [ ] **1 tub**
- [ ] **4 nesting containers**
- [ ] **up to 4 half-liter bottles**
- [ ] **up to 4 bottle lids**
- [ ] **up to 6 clear cups**

Store these chapter-specific items together in a designated place. They require about 1/2 square foot of dedicated space. General classroom materials (like scissors and tape) are also listed below when used, while others (like pencil and paper) are always assumed.

scrambled movie frames (pour 1)

Photocopy and cut. Arrange these frames in any random order; **paper clip** together.

40 rings (pour 3)

Cut 10 empty **toilet-tissue rolls** into 4 rings each. Begin each cut by poking pointed scissors through the roll. This makes a total of 40 rings. Number each ring on at least 3 sides with **colored markers**. Create color-coded sets of ones, tens, fives and twos:

ones / black: 1, 2, 3, 4, 5, 6, 7, 8, 9, 10.
tens / brown: 10, 20, 30, ..., 80, 90, 100.
fives / blue: 5, 10, 15, ..., 40, 45, 50.
twos / red: 2, 4, 6, 8, 10, 12, 14, 16, 18, 20.

Tie a **paper clip** to each end of about 1 foot of **string**. String like colors together, fixing the first and last ring to the paper clips. Store in a **gallon storage jug**. Paper label provided on next page.

plastic, mirror, metal (pour 5)

Store 3 small, flat items made from these materials in a **plastic sandwich bag**. We used a **canning lid**, a **small yogurt lid**, and a **small glass mirror**.

Masking tape label: *pour 5*.

fraction spoons (pour 6)

Label 20 **craft sticks** with a fraction at one end (on both sides). Use a **fine-tipped marker**:

$1/2, 1/2$; $1/3, 1/3, 1/3$; $1/4, 1/4, 1/4, 1/4$;
$1/5, 1/5, 1/5, 1/5, 1/5$; $1/6, 1/6, 1/6, 1/6, 1/6, 1/6$.

Rubber-band them together.

double lid (pour 8)

Slice the "roofs" off two **liter-bottle lids**, leaving two threaded tubes. Do this by hand with a **hacksaw** (or an electric band saw.) Tightly screw each lid onto a liter bottle to provide a safe grip. Cut into the top edge of each cap by only the thickness of the blade. The bottle's rim helps guide the saw blade, keeping you from shaving off too much. When both caps are cleanly cut, join them cut edge to cut edge with a single layer of masking tape. Then overlay with a long strip of plastic **electrical tape** stretched and wound very tightly several times to make a solid connection. (Note: Without a masking-tape base, the tight electrical tape may slip over time, forcing the caps off-center.)

Such connectors also come ready-made. Look for "tornado tubes" sold in science supply catalogs. Or you might make a permanent hourglass by joining liter bottles directly with electrical tape over a masking tape base.

double lid (pour 8)

fraction spoons (pour 6)

plastic, mirror, metal (pour 5)

40 rings (pour 3)

scrambled movie frames (pour 1)

POUR 3
40 rings
(4 sets of 10)

Apply to a gallon storage jug. Use
clear packaging tape.

A / POUR
special materials

CHAPTER SIGN: Glue to a 4 x 6 inch index card, and fold in half.
Stand this sign in the space where you store special materials for this
chapter. Or cut this sign in half, and glue both pieces to a grocery
bag that has been cut to size, or to a box.

Preparation: 1 copy

TEACHING NOTES

Purpose

To practice a cleanup procedure for gathering and storing lentils. To arrange a series of pouring pictures in the correct sequence.

Introduction

◗ Pour a liter of lentils into an empty Job Box, then scoop them back into the bottle. Emphasize these techniques as you scoop and funnel:

• Hold the scoop flat, or at a very shallow angle, to efficiently capture lentils. Avoid steep angles.

• Grip the funnel at its very bottom, so your hand can grip the top of the bottle as well. This leaves your other hand free for pouring.

◗ Recap the liter of lentils and return it to its designated place. Also put back the scoop and funnel. Every job ends just where it begins, with all materials in their proper storage places.

Focus

◆ What is the Job Card asking you to do? *Pour the bottle of lentils into a mound, then scoop and funnel them back into the bottle. Anything else? Put the scrambled movie frames in order.*

◆ As you pour lentils…

How does the bottom of the box change?
How does the inside of the bottle change?

◆ Can you pour lentils back into the bottle without spilling even one?

◆ Do all the lentils fit back into the bottle? (If students started with a full bottle, firmly packed, and refill the bottle without shaking it, the lentils may not all fit! Not until they settle its contents by gently shaking the bottle or "tickling its ribs.")

◆ Does the funnel work upside down? Why?

◆ Can you correctly arrange these Movie Frames, in left-to-right reading order? (Flip the pages with thumb and forefinger to create the illusion of motion. The out-of-sequence pictures seem to move helter-skelter, but only because they are not properly sequenced.)

Checkpoint

◆ Show me how the scoop works.

◆ Show me how to hold the funnel with one hand and scoop with the other.

◆ Show me that these Movie Frames are arranged in the correct order. (Older children might have the dexterity to flip a "moving picture.")

◆ Draw and write about what you learned.

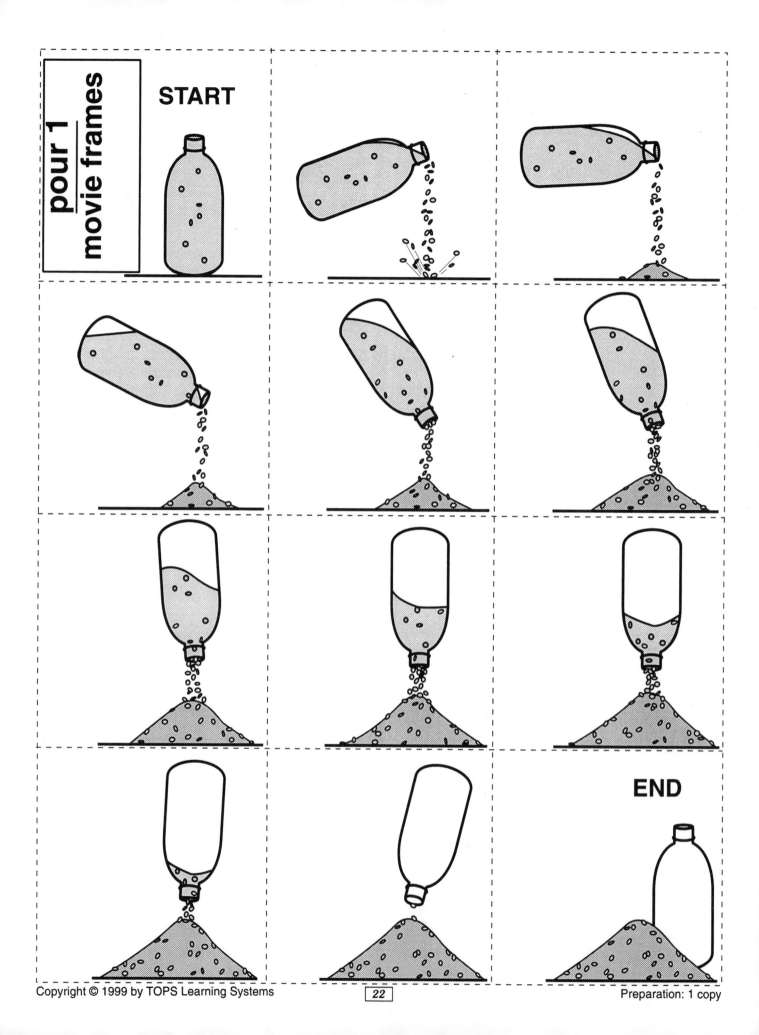

Preparation: 1 copy

TEACHING NOTES

Purpose

To draw "before," "during" and "after" frames of a sequential process.

Introduction

➧ Invert a full liter of lentils over an empty Job Box until about half pour out. Stop the flow with a bottle lid to freeze the action.

➧ While holding the bottle upside-down, ask volunteers to describe exactly what they see…

- How full is the bottle? How do the lentils look inside?
- How do the lentils in the box look? What shape do they take?
- Remember to look carefully at each thing you draw. There is only one part you must draw from memory. What is it? *Lentils falling from the bottle to the box.*

Focus

◆ What is the Job Card asking you to do? And then what?

◆ Can you pour out just half the lentils? (Students may find it helpful to cap the bottle while they study this stage.)

◆ Can you draw the middle (during) picture first, showing how lentils look in the bottle and the box? Draw what you *really* see, not what you *think* you see.

Checkpoint

Let me see your drawing. How does it show changes…

- inside the bottle?
- inside the box?

More

Optical Illusion: Stare at the point where the lentils are landing, for the full duration of the pour (12-15 seconds). After all motion stops, the lentils will appear to move in the opposite direction in slow motion, and the mound will seem to contract! (This illusion is likely beyond the capacity of most young children due to lack of sustained concentration.)

A/2 You need...

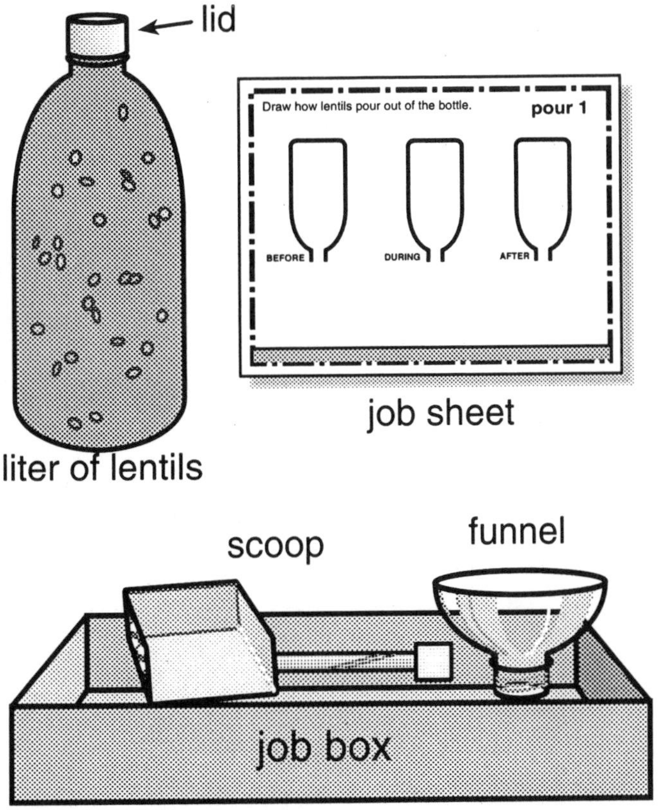
lid
Draw how lentils pour out of the bottle.
pour 1
BEFORE
DURING
AFTER
job sheet
liter of lentils
scoop
funnel
job box

pour 2

Pour and stop halfway.
Draw how lentils pour out of the bottle.
pour 1
BEFORE
DURING
Draw what you see.

pour 2

Draw how lentils pour out of the bottle.

BEFORE|

DURING|

AFTER|

Job Sheet: 1 copy per student

TEACHING NOTES

Purpose

To construct a tower of numbered rings filled with lentils. To practice counting by ones, twos, fives, and tens, both forward and backward.

Introduction

Let's see if I can stack up these 10 rings into a tower. (It will probably topple over before reaching that number.)

Ooops! Perhaps I can do better by first filling each ring with lentils. (Filling them one at a time is easier than trying to hold several together as you pour lentils in.)

Focus

◆ Which set of rings would you like to stack?

◆ Can you build a tower of rings, with the smallest number on the bottom and working up?

◆ Can you count forward as you build each tower?

◆ Can you count backward as you take the rings off again?

Checkpoint

◆ Let me hear you count forward from __ to __.

◆ Let me hear you count backward from __ to __.

◆ If I take away these __ rings, how many are left?

◆ Are the rings of this tower in the correct order?

◆ Beginning with the lowest number in the tower, write forward by 1's, 2's, 5's, 10s.

◆ Beginning with the highest number in the tower, write backward by 1's, 2's, 5's, 10s.

A/3 **You need...**

jug of rings

pour 3
40 rings
(4 sets of 10)

job box with **1** liter of lentils

pour 3 👤 or 👤👤

Build a tower.

Make it shower.

TEACHING NOTES

Purpose

To understand that 1 cup of lentils can spread outward or fill upward, depending on the width of the container.

Introduction

♦ Rubberband an empty one liter bottle at about 3/4 of its capacity. Fill a small bottle to the top with lentils. Stand the bottles side by side.

♦ Ask your class to predict how high the lentils will reach when you pour them from the narrow to the wide bottle. Who thinks the lentils will...

　...reach HIGHER than this rubber band?
　...reach EVEN with this rubber band?
　...reach LOWER than this rubber band?

A prediction is an intelligent guess based on what you already know. Ask volunteers to explain why they predicted one way and not another.

♦ Now pour the lentils into the liter bottle. Notice that they don't reach the rubber band. Roll the rubber band down to their actual level. Using a funnel, pour the lentils back and forth several times, noting that they always reach the same higher mark in the narrow bottle; the same lower mark in the wide one. No lentils get lost or found in the transfer.

♦ Sketch both bottles on your blackboard and draw how the lentils look inside. Use light shading to represent the whole mass of lentils, with perhaps an occasional circle to suggest individual seeds. Don't try to draw every lentil – there are too many! Suggest the top surface in each bottle with a solid line.

Focus

◆ How much (what volume) will you pour into each nesting container? *The same amount: as much as the narrow container will hold.*

◆ Do these nesting containers have the same height? The same width? *All heights are the same; all widths are different.*

◆ How will you show the lentils in your drawings? *With light shading and an occasional circle. Use a straight line for the top surface.*

Checkpoint

◆ Let me see your drawings.

◆ Why do the lentils reach LOWEST in this container? *It is the widest; they spread out more.*

◆ Why do the lentils reach HIGHEST in this container? *It is the narrowest; they pile up higher.*

A/4   You need...

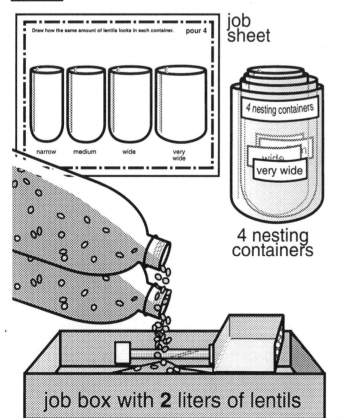

job sheet

4 nesting containers

job box with **2** liters of lentils

pour 4

Pour the same amount.

I will put this much in each container.

Draw how the SAME volume changes shape.

pour 4

Draw how equal amounts look different.

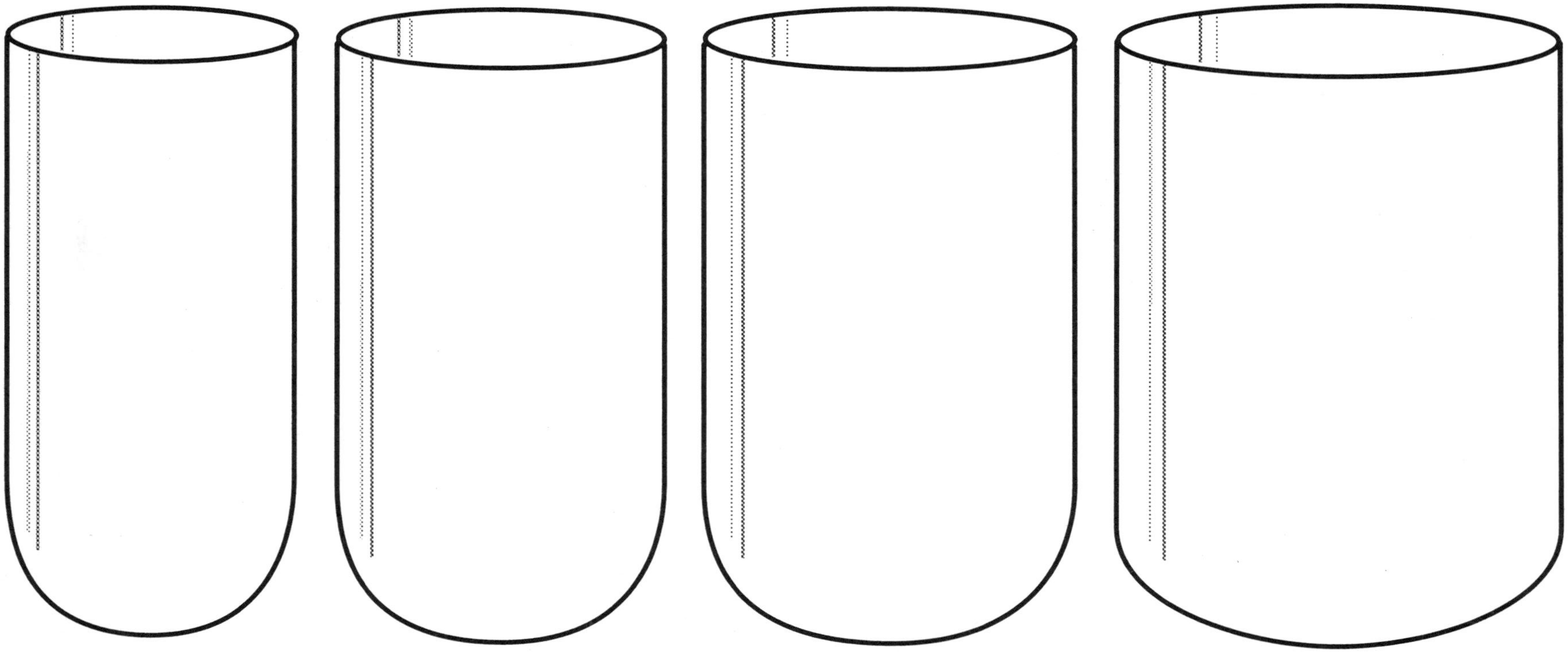

very
wide

wide

medium

narrow

Job Sheet: 1 copy per student

TEACHING NOTES

Purpose

To experience lentils with eyes, hands and ears. To use rich vocabulary to describe them.

Introduction

Eye, Hand and Ear are telling each other about lentil science. Who can think of words that each character might use to talk about lentils?

Allow students to dip into a box of lentils with their hands for inspiration. List student responses on your blackboard under these headings:

- Eye: Lentils LOOK…
  small, flat, dusty, round, brown, broken;
  like clam shells;
  like bald headed babies;
  like little hocky pucks.
- Hand: Lentils FEEL…
  hard, cool, dry, smooth, slippery, squishy;
  like lizard scales;
  like small coins;
  like tiny pieces of soap.
- Ear: Lentils SOUND…
  loud, pounding, tapping, soft, whispering;
  like tiny drums;
  like a hailstorm;
  like light rain.

Focus

◆ How do lentils look on the mirror?

◆ How does your hand feel when it gets buried?

◆ How does the sound of the lentils change as they strike each surface?

Checkpoint

◆ Tell me a word that describes how lentils look/ feel / sound.

◆ Finish the sentences: Lentils look / feel / sound like...

◆ Show your written work. (Beginning writers might copy a few words from the Job Sheet into each category. Urge more capable students to search a dictionary for new words on their own.)

More

We have been using 3 out of 5 senses. Which two are missing? (We are not *tasting* the lentils because they are dirty. But they have a delicious nutty flavor when washed and cooked. We are not *smelling* the lentils either, though they do have a faint musty odor when smelled up close.)

Caution

Please: No lentils up the nose or in the ears. This is dangerous. They could get stuck.

A/5 You need...

plastic, mirror, metal

job sheet

job box with **1** liter of lentils

pour 5   **🧍 or 🧍🧍**

Look.   Touch.   Listen.

mirror   plastic   metal

Write!

 **How lentils LOOK:**

 **How lentils FEEL:**

 **How lentils SOUND:**

---

whispering	tapping	round	like small coins	like pebbles
smooth	cool	hard	like bald-headed babies	like clam shells
loud	brown	pounding	like tiny drums	like hocky pucks
small	soft	squishy	like scales	like little bells
slippery	broken	flat	like rain	like lentils

   Job Sheet: 1 copy per student

TEACHING NOTES

Purpose

To divide a liter of lentils into equal portions. To understand the relationship between equal parts and the whole.

Introduction

♦ Hold up a small bottle of lentils. (Notice that it is smaller than the liter bottle.) Imagine it is filled with chocolate pudding. Stick 3 Fraction Spoons into the top, labeled $1/3$, $1/3$ and $1/3$. This means the pudding must be equally divided between 3 people.

• Pour *all* of the pudding into 3 cups in an empty Job Box. Pour back and forth between cups, as necessary, to equalize the 3 portions. Congratulate yourself on keeping the floor of your box clean.

• Invite 2 students to join you for "dessert." Put a Fraction Spoon into each cup: $1/3$ for you, $1/3$ for you, $1/3$ for me. Are the servings even? Pretend to eat.

• Pour all cups back into the bottle, counting as you go: 1 third, 2 thirds, 3 thirds fill the whole. Stick the Fraction Spoons back into the bottle.

♦ Repeat this pattern with the 4 Fraction Spoons each labeled $1/4$. Ask volunteers to assist, where possible, but don't let the pace slow so much that students lose interest.

Focus

♦ Into how many equal parts will you divide the whole bottle: 2, 3, 4, 5, or 6? Which Fraction Spoons will you use?

♦ What will you do if "pudding" spills on the floor of the box? *Put it back into the bottle; it must all end up divided among the cups.*

♦ After all the "pudding" is poured, look at the cups. Which one would you choose? (If it's hard to decide, they are nearly equal.)

♦ Can you count the parts as you pour them back into the whole? (Example: 1 sixth, 2 sixths, 3 sixths, 4 sixths, 5 sixths, 6 sixths fills 1 whole.)

Checkpoint

♦ Please arrange 5 equal cups of pudding with a Fraction Spoon in each one. Tell me when you are ready: I would like to watch you pour and count them back into the whole.

♦ Draw and write about what you did and what you learned.

More

Fold a sheet of paper into 4 equal parts. Label each part and count them. Repeat for 8 equal parts; 16 equal parts.

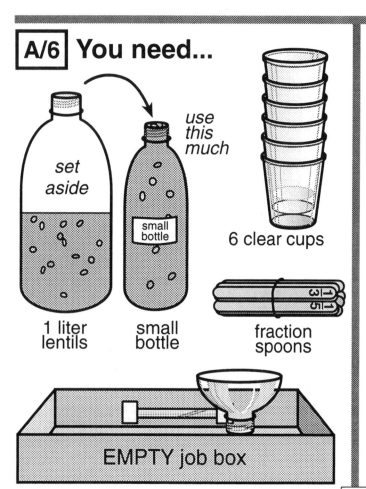

A/6 You need...

set aside

1 liter lentils

use this much

small bottle

small bottle

6 clear cups

fraction spoons

EMPTY job box

pour 6

Divide the small bottle into EQUAL parts.

small bottle

$1/3$ $1/3$ $1/3$ $1/3$

Label each part.

☞ Keep all lentils in containers.

TEACHING NOTES

Purpose

To practice accurate representational drawing through careful observation.

Introduction

▶ Children (adults too) often set to work drawing a mental picture of what they *think* they see, without really looking at the object at all. You can begin to break this habit by modeling how to draw a half-filled bottle of lentils standing on your table. Observe it in 5 stages, finger tracing each part in the air before you draw it on your blackboard.

• Look at the bottle top and draw it.
• Look at the shoulders and draw them.
• Look at the sides and draw them. (Simplify any ridges.)
• Look at the bottom and draw it. (Simplify by eliminating indentations.)
• Look at the lentils inside and draw them. (Represent with light shading and a few small circles. Represent the top surface of the lentils with a line.)

▶ Sketch the top of the lentils at a slant. Encourage your class to correct this "mistake," by helping you notice that lentils rest level in a shaken bottle.

Focus

◆ What steps do you need to do as you set up the "still life" in this Job Card? (Students should fill each small bottle half full, close with a lid, and shake level before setting in place.)

◆ Can you draw how the lentils look inside each bottle on this Job Sheet?

◆ If you shake the bottle, do the lentils always settle level? *Yes.* (The lentils rest horizontal no matter how the bottle is turned. Only keen observers will draw a horizontal surface in the tilted bottle without prompting.)

Checkpoint

◆ Are the lentils represented with light shading?

◆ Do lentils in the inverted bottle rest higher than lentils in the right-side-up bottle? *Yes.* (The inverted bottle is narrower at the bottom, so the lentils rest higher. Students easily miss this distinction.)

◆ Is the surface of the lentils level (horizontal) in each drawing no matter how the bottle is tilted or

A/7 **You need...**

lids

4 small bottles

a liter of lentils

tub

job sheet

EMPTY job box

pour 7 🧍 or 🧍🧍

Set up a still life.

Half full. Capped. Shaken level.

Draw the lentils.

Draw the lentils at rest. pour 7

UP   DOWN   SIDEWAYS   TILTED

Draw the lentils at rest.

pour 7

TILTED

SIDEWAYS

DOWN

UP

32

Job Sheet: 1 copy per student

TEACHING NOTES

Purpose

To explore the properties of an hourglass. To develop an ability to estimate the passing of time in seconds.

Introduction

▶ Practice counting 5-second time intervals with your class: a thousand one, a thousand two, a thousand three, a thousand four, a thousand five. Use a large wall clock to establish timing and rhythm. Notice that 5 seconds is the time needed for the second hand to pass from one large number to the next on the face of the clock. Work up to 10 and 15 second intervals.

▶ Ask the children to watch as you make an hourglass: Fill a liter bottle to the top without "tickling" the sides or settling the contents in any way. Screw the Double Lid on top, then connect a second empty bottle inverted over that.

▶ Wonder out loud if this hourglass can measure time in seconds. Allow children to explore this on their own.

Focus

◆ Can you time how long it takes the hourglass to run down?

◆ Can you make the hourglass run slower? *Lean it a little to one side.*

◆ How many times can you stand up and sit down during 1 cycle of the hourglass?

Checkpoint

◆ Let me hear you time the hourglass by counting seconds. (It typically takes from 12 to 15 seconds to empty when held vertically.)

◆ What else did you learn about this hourglass? Can you draw and write about it?

◆ Let me see your drawings.

More

◆ What happens if you fill the lower bottle with densely packed lentils? *When I tip the hourglass, the bottom fills up before the top fully empties.*

◆ Can you make an hourglass with 2 bottles of different size? How full should you fill it? (Provide a half liter bottle with compatible threading.)

◆ Can you draw a picture of how the lentils move from top to bottom?

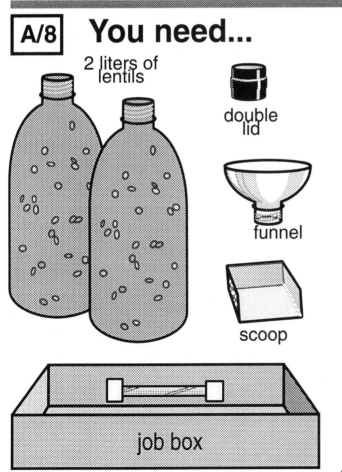

A/8 **You need...**

2 liters of lentils

double lid

funnel

scoop

job box

pour 8 🧍 or 🧍🧍

Can you make an hourglass?

How many seconds will it run?

... a thousand 5, a thousand 6...

Draw lentils!

TEACHING NOTES

Purpose
To provide students with an approved way to pursue their own ideas about pouring lentils. To encourage creativity.

Introduction
Display this card whenever you want to do this suggested activity or experiments you design yourself.

Focus
◆ What do you want to study about pouring?
◆ Are these the materials you need?

Checkpoint
Report in words and pictures:
• What you did.
• What you learned.
• Questions you may still have.

Special Note
Use the opposite page to record especially clever ideas. Use them to inspire students that follow to be even more creative and inventive.

Send TOPS these ideas, too. We may include some in a future edition of this book, as sparks to ignite the imaginations of future young scientists.

Better yet, let your students address their own ideas to the world! Suggest that they write a report, and send it to us at the address on the title page. We'll send an affirming "thank you" note.

A/9 **You need...**
these BASICS:

job box

☞ Ask your teacher for other items.
Tell why you need them.

pour 9 🧍 or 🧍🧍

On your own.

My idea:
A holder for
the funnel

Both hands free!

B / SEARCH

In this chapter: Look for pintos, reds beans and pennies among the lentils. Pretend they're rabbits, and you're the fox. Screen the lentils in systematic ways to separate large from small. Distinguish size and texture by touch alone. Fish for paper clips with magnetic bait. Bulldoze for BB's, and make them dance!

job card **1**	lost and found pintos	38
job card **2**	hide the rabbits	39
supply	story paper	40
job card **3**	blind search	41
job sheet	finger math	42
job card **4**	screen beans by size	43
job card **5**	fish for paper clips	44
job card **6**	bulldoze for paper clips	45
job card **7**	what can you do with 30 BB's?	46
job card **8**	on your own (sharks!)	47

Basic Materials: Quantities define maximums needed to support any one Job Card in this chapter. Store *high-quantity basics* (Job Boxes, liters of lentils, bottle lids, scoops, funnels) on and under a table or counter. Store *low-quantity basics* near the "basics" sign (see page 49) or in a "basics" box. Consult our Glossary on pages 6-10 for a full description of these items. See the next page for additional special materials used in this chapter.

☐ **1 job box**
☐ **1 liter of lentils**
☐ **1 scoop**

☐ **1 paper plate**
☐ **story paper**
☐ **2 tubs**

Store these chapter-specific items together in a designated place. They require about 1 square foot of dedicated space. General classroom materials (like scissors and tape) are also listed below when used, while others (like pencil and paper) are always assumed.

~75 pintos (search 1)

Buy a **bag of pinto beans**, or sort them out of the bean mix also used in this chapter. Fill two **film cans** to capacity and snap on the lids. Each canister will hold approximately 75 pintos.

Paper labels provided on next page.

screen (search 1, 2, 4)

Purchase **1/4 inch grid hardware cloth** (wire screen) from a hardware store. Other mesh sizes will *not* work. Use **wire cutters** to cut a full 6 x 6 inch piece, *plus* an extra fringe of half squares all around the perimeter. Cover exposed edges with 3/4 inch plastic **electrical tape**, folded evenly over the wire and stuck to itself.

Trim the tape at each corner. Bend up a half inch lip all around to create a shallow, square dish.

10 pennies, 10 pintos, 10 reds (search 2)

Sort red beans and pintos from the **bean mix** used in this chapter. Place these seeds and the pennies in the same **film can**, and snap on the lid.

Paper label provided.

eye covers (search 3)

Join two **baby food jar lids** and two **medium-to-large rubber bands** with long strips of masking tape in a goggle configuration.

Keep the lids separated by the width of the masking tape. Cover the exposed sticky nose bridge with a short vertical piece of tape. Fit for comfort to child's head size with a rubber band around each ear. Add or modify rubber bands as necessary. (Goggles should fit tightly on adult heads.)

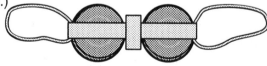

5 marbles, 5 garbanzos, 5 limas (search 3)

Sort garbanzos and giant limas from the **bean mix** used in this chapter. Pack a **film can**, first with the 5 marbles, then the 5 limas, then the 5 garbanzos. Snap on the lid.

Paper label provided.

bean mix (search 4)

Purchase two 1-pound bags of **15- or 16-variety soup mix**. Try to find a mix that contains dried whole peas, a very useful seed not available in all mixes. (Try Western Family brand in western states.) Fill a **quart jar** about 3/4 full, and screw on the **lid**.

Masking tape label: *bean mix.*

4 magnets (search 5, 6, 7, 8)

Purchase rectangular, **ceramic refrigerator magnets** from an electronics store, science supply catalog, or directly from TOPS. Ours measure 1 inch x 3/4 inch x 1/8 inch.

Wrap a strip of masking tape around three of these magnets, and label the poles N for north-facing, S for south-facing. Stand a new **brad** on its head at the center of the north pole on the fourth magnet. Wrap with masking tape to hold it in place. Attach a short length of **string** with more tape, and label the poles. Any number of magnets can now hang from the string with poles oriented up and down. Interesting pendulum interactions are possible, because the brad keeps repelling magnets from easily flipping over.

20 paper clips (search 5-6)

Place in a **film can**, and snap on the lid. Paper label provided.

wood blocks (search 6, 7, 8)

Cut two **2 x 4's** to about 15 inches, the width of your Job Box.

30 BB's, 1 bottle cap (search 7, 8)

Purchase BB shot from the camera and ammunition case of your local drug or variety store. Place in a **film can**, along with an unbent twist-off **bottle cap**, and snap on the lid.

Paper label provided.

circle pattern (search 7, 8)

Cut a flat, translucent, 1 x 2 inch piece of plastic from a **"clamshell" food container or deli lid**. Round all sharp corners. **Paper-punch** a hole near one end.

Tag with a masking tape label: *circle pattern.*

paper clip (search 7)

circle pattern (search 7, 8)
30 BB's, 1 bottle cap (search 7, 8)
wood blocks (search 6, 7, 8)
20 paper clips (search 5, 6)
4 magnets (search 5, 6, 7, 8)
bean mix (search 4)
5 marbles, 5 garbanzos, 5 limas (search 3)
eye covers (search 3)
10 pennies, 10 pintos, 10 reds (search 2)
screen (search 1, 2, 4)
~75 pintos (search 1)

B / SEARCH
special materials

CHAPTER SIGN: Glue to a 4 x 6 inch index card, and fold in half. Stand this sign in the space where you store special materials for this chapter. Or cut this sign in half, and glue both pieces to a grocery bag that has been cut to size, or to a box.

Apply each label to a film canister with clear packaging tape.

search **1**

~ 75 pintos

search **1**

~ 75 pintos

search **2**

10 pennies
10 pintos
10 reds

search **3**

5 marbles
5 garbanzos
5 giant limas

search **5-6**

20 paper clips

search **7**

30 BB's
1 bottle cap

Preparation: 1 copy

TEACHING NOTES

Purpose

To search for pinto beans among the lentils in a systematic manner. To practice concentration and observation skills.

Introduction

▶ With the children watching, take 10 pinto beans from the film can and mix them into a liter of lentils in the Job Box. Invite volunteers, in groups of 2 or 3, to look for them. Keep track of how many beans are found and how many are still lost. (Children have excellent near vision. They may be better at finding the pintos than you are. Good lighting is always helpful.)

▶ When only 1 or 2 pintos remain, demonstrate how to find these holdouts by sifting the lentils with a screen. Model how to do this systematically, gathering unscreened lentils from one side of the box, and discarding screened lentils on the other side of the box.

Focus

◆ Can you lose a film-can-full of pinto beans in the lentils and find them all again?

◆ Can you use the screen to help you search?

Checkpoint

◆ How does the screen help you find the pintos? *The small lentils fall through the screen, while the larger pintos are left on top.*

◆ How should you search so you don't have to look through the same lentils twice? *Be systematic: scrape all the "unsearched" lentils to one side of the box, then isolate the "searched" lentils (those you have checked carefully or screened), on the other side of the box.*

◆ How do you know when you have found all the pintos? *You can't be sure unless you count what you started with. But if you can pack the cylinder full again, you know you're close.*

◆ Draw and write about what you did and what you learned.

More

◆ How many pintos does the film can hold? *Group and count by fives. There will be 70 to 75 pintos, depending on size.*

◆ Can you lose and find 2 film cans of pintos? Does this take twice as much time?

◆ How long does it take to find 1 film can of pintos in 2 liters of lentils?

B/1 You need...

screen

about 75 pintos

search 1

scoop

1 liter of lentils

search 1   🧍 or 🧍🧍

**Mix in all the pinto beans.
Find them again.**

TEACHING NOTES

Purpose

To search for 30 "rabbits" of 3 different "species" in a lentil habitat.

Introduction

Mix all 30 "rabbits" into the lentils, then ask volunteer "foxes" to find them and put them on a paper plate. Stop the search after about 20 in all have been found.

• Ask how many of each kind of "rabbit" are still hiding. (This is an instant math lesson on subtracting from 10.)

• Notice which rabbits were "eaten" and which still survive. (Pintos hide better than reds because of camouflage: they blend in better with the lentil background. Pennies hide best of all because they "burrow underground." The last penny or pinto may require screening to locate.)

Focus

◆ Can you mix these rabbits into the lentils and find them again? Which do you think you will find most easily? Why?

◆ How will you know when you find all the rabbits? (There will be 10 each, or 30 altogether.)

Checkpoint

◆ Did some rabbits hide better than others? Why?

◆ What kind of rabbit would you like to be if you lived in a lentil habitat? Why?

◆ What kind of a rabbit would you like to be if you lived in dark red gravel? Why?

◆ Did you use the screen? Can any rabbits escape that? Why?

◆ Write a story about rabbits and foxes. Illustrate your work with a drawing.

More

◆ Can you find penny rabbits by the sound that they make? (Yes. Scoop up the lentils and pour them back into the Job Box from a foot or more high. Listen carefully, and you can hear the characteristic metallic clink of a penny that is distinct from the whisper of falling lentils.)

◆ Any snow outside? Scatter lima rabbits and pinto rabbits outside. Who best survives the hunt?

◆ Make up a story about small animals living in a wild place. Draw pictures to go with it.

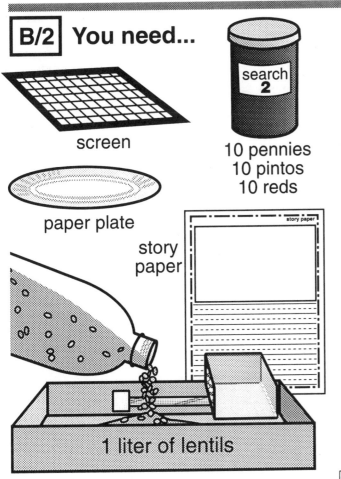

B/2 You need...

screen

search 2

10 pennies
10 pintos
10 reds

paper plate

story paper

story paper

1 liter of lentils

search 2   👤 or 👥👥

Hide 30 rabbits.

10 penny rabbits

10 red rabbits

10 pinto rabbits

Who survives the hungry fox?

Supply: several copies per student

TEACHING NOTES

Purpose

To search for seeds and marbles among the lentils with eyes covered, using only the sense of touch.

Introduction

Stand before a clean table with just 1 pinto bean in the middle. Ask a volunteer to move the pinto anywhere on this table while you wear the Eye Covers. Model how to do a systematic blind search.

• What sense did I use to find the bean?

• How did I make sure I searched the whole table?

Focus

◆ Can you mix in the marbles, garbanzos and limas, then find them with your eyes closed?

◆ How do marbles, garbanzos and giant limas feel different from each other?

Checkpoint

◆ How did each object feel? *The marbles feel large, round and cold; the garbanzos feel rough and pointy; the giant limas feel smooth and cool.*

◆ Which objects were easiest to find? Hardest? *Marbles are easiest; garbanzos are more difficult; limas are hardest, requiring focused attention on the sense of touch.*

◆ Let me see your Job Sheet. Did you complete the finger math?

More

◆ Study animals that use a sense of touch to find food. Do they also use a sense of smell? Examples: jellyfish, moles.

◆ Write story problems for classmates to solve. Example: Lon bakes a dozen cookies and gives away 3. How many remain?

Special Note

Putting these objects back in the film can is an interesting puzzle. The largest marbles must be packed first, followed by the smaller limas and garbanzos. If the marbles are packed last, they will not fit! (Is there a lesson here about life? Do all the little projects leave no time for big ones?)

B/3 You need...

paper plate

eye covers

search 3

5 marbles
5 garbanzos
5 giant limas

finger math: FIVES search 3

| marbles found | garbanzos found | giant limas found |
| still lost: | still lost: | still lost: |

TENS

pinto beans found	red beans found	pennies found
still lost:	still lost:	still lost:
BB's found	paper clips found	pinto beans found
still lost:	still lost:	still lost:

job sheet

job box with **1** liter of lentils

search 3

Blind Search

Search 3

Do finger math.

finger math

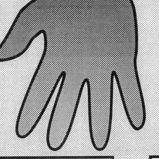

HIDE FIVE

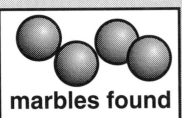

marbles found
_____

still
lost:

garbanzos found
_____

still
lost:

giant limas found
_____

still
lost:

HIDE TEN

pinto beans found
_____

still
lost:

red beans found
_____

still
lost:

pennies found
_____

still
lost:

BB's found
_____

still
lost:

paper clips found
_____

still
lost:

pinto beans found
_____

still
lost:

Job Sheet: 1 copy per student

TEACHING NOTES

Purpose

To sort mixed beans by size, using a screen.

Introduction

Demonstrate a sorting procedure. Pour beans into the screen and shake it over a paper plate. Put the little beans that fall through into one tub; the bigger ones that remain on top into the other. All operations, of course, must take place inside an EMPTY box.

Empty box? Yes, EMPTY. Caution your students never to pour these beans into a box of lentils. It takes too much work to separate them back out! (If this should happen, screen out larger seeds, then require the offender to separate *most* lentils from the remaining small seeds by hand. Another stragegy is to blend your own mix composed only of large seeds and small lentils. This less diverse mixture can be separated solely by screening.)

Focus

◆ (Level 1) Can you find the big seeds?

◆ (Level 2) Can you sort the big seeds into 1 tub, and the little seeds into the other? (When children are demonstrating organization skills and accuracy, not overfilling the screen or spilling beans over the top, suggest level 3.)

◆ (Level 3) Can you sort the seeds into 3 tubs: large, medium and small? (First sort into large and small grades. Then filter both tubs again. Any large seeds that pass through on second screening, and any small seeds that don't pass through on second screening, can be grouped into a "sometimes-pass-through" medium grade.

Checkpoint

◆ How did you use the screen to separate the large seeds from the small ones?

◆ Show me a kind of seed that…
…almost *always* passes through the screen.
…almost *never* passes through the screen.
…*sometimes* passes through the screen.
(Students don't need to know the names of seeds to point out representative types: SMALL = barley, lentils, split peas; LARGE = lima, kidney, pinto, garbanzo; MEDIUM = pink, white, red, black, whole peas.)

More

Pour the sorted beans back into the jar. Before remixing, draw how the small beans remain separated from the large with a distinct interface.

B/4 You need...

screen

bean mix

NEVER mix with lentils.

2 tubs

paper plate

EMPTY job box

search 4 🚶

Screen mixed beans by size.

small

large

How did you do it? ✏️

TEACHING NOTES

Purpose

To do a systematic search for paper clips using a magnet.

Introduction

Demonstrate how to "fish" for paper clips, by passing the magnet just above the surface of the lentils. Scan back and forth in a systematic way. Using reasonably strong magnets, the paper clips will literally jump out of the lentils!

Focus

◆ Can you stock 20 fish in Lentil Lake and catch them all?

Checkpoint

◆ Can you catch a fish without touching the lentils with the magnet?

◆ How else can you use the magnet to find the paper clips? *Drag the magnet through the lentils.*

◆ If you hide 20 fish, and catch 16, how many remain?

◆ Let me see your story problems.

More

Crazy Pendulum: Put a block of 2 magnets (south pole up) on your table. Suspend the 2 remaining magnets by the string over these table magnets. The suspended magnets will repel in fascinating ways.

B/5	**You need...**

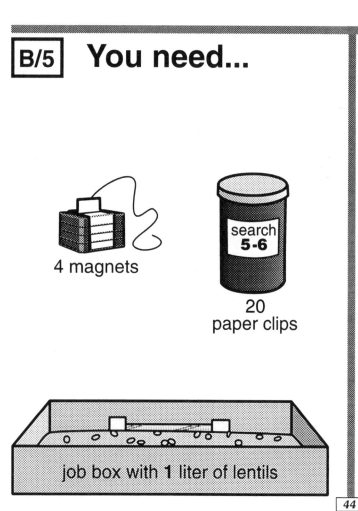

4 magnets

search **5-6**

20 paper clips

job box with **1** liter of lentils

search 5

Fish for paper clips.

I hid 20 "fish."

Lentil Lake has

Write story problems.

TEACHING NOTES

Purpose
To invent stories and games. To have fun.

Introduction
Give the Job Box a quick shake to evenly distribute the liter of lentils across the bottom. Then set it on two 2 x 4 blocks so you can freely pass 2 magnets underneath. Demonstrate how to pinch the tape that sticks out from one of the magnets between closed straight fingers, either palm up or palm down.

Focus
Pretend you are…

- Driving a bulldozer: Build a road. Clear a parking lot.

- Searching for buried treasure: Find all 20 paper clips.

- Driving a race car: Push back the lentils to create a paper clip obstacle course in the middle of the job box. Steer your magnetic car through this course without picking up a single paper clip.

Checkpoint
Draw a picture and write a story about what you did.

| B/6 | **You need...** |

story paper

20 paper clips — search 5-6

4 magnets

job box with **1** liter of lentils

wood blocks

search 6

Bulldoze for paper clips.

2 magnets

2 magnets

Write a story.

TEACHING NOTES

Purpose

To explore, invent, and engage the imagination. To express ideas using creative drawings and rich language.

Introduction

Pass around a bottle cap packed with one layer of BB's. They form a square of 4 in the middle, surrounded by an inner circle of 10 and an outer circle of 16. Pressing them into the bottle cap is a quick and easy way to know whether you have found all 30 BB's!

Focus

Think of something wonderful that you can do with BB's, magnets, and a Job Box of lentils…

• Clear away an empty space inside the lentil box. Make the BB's "dance" by moving magnets underneath the box.

• Make 4 BB's dance together. Use the Circle Pattern to trace little circles that show different patterns and combinations of 4. Repeat with other numbers of BB's.

• Hide BB's in the lentils. Search with magnets from above or below.

• Burrow through the lentils with BB or paper clip "prairie dogs."

• Fashion a paper corral. Use BB's as "sheepdogs" to round up lentil "sheep."

• Bend a paper clip into an open pen. Try to get BB's into the pen before it turns the wrong way.

• Draw a maze on paper. "Drive" a BB through it without touching the lines.

• What else?

Checkpoint

◆ Report what you did and what you learned.

◆ Draw a picture

◆ Write a story.

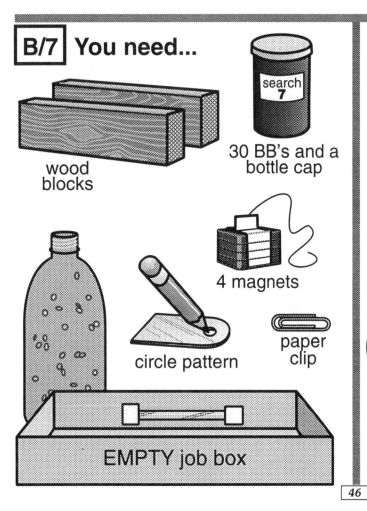

B/7 You need...

wood blocks

30 BB's and a bottle cap

4 magnets

circle pattern

paper clip

EMPTY job box

search 7

Start with 30 BB's in a bottle cap...

What can you do with them?

4 10 16 — 30 BB's

IDEAS:
• *dance* • **draw** • hide • search • *fish* •
• burrow • bulldoze • ***round up*** • drive •

TEACHING NOTES

Purpose
To provide students with an approved way to pursue their own ideas about searching through lentils.

Introduction
Display this card when you want to do the suggested activity or experiments that you design yourself.

Focus
◆ What do you want to study about searching?
◆ Are these the materials you need?

Checkpoint
Report in words and pictures:
• What you did.
• What you learned.
• Questions you may still have.

Special Note
Applaud creativity: Record some of your students' most amazing ideas on the opposite page.

B/8 **You may use...**
these BASICS:

search 7

job box with **1** liter of lentils

☞ Ask your teacher for other items. Tell why you need them.

search 8   👤 or 👥

On your own.

My idea:
Big fish!

C / COMPARE

In this chapter: Ten containers of different sizes, each with a different symbol. All are shapes that kids can draw: a sun, moon, heart, star…. The moon holds less than the star, but more than the sun. The moon equals 4 hearts. Is this poetry or is it math? Gently, creatively, children move from the concrete world of lentils and containers to the abstract world of equality and inequality. Ideas about quantity and number emerge slowly and naturally, through experience, the best way to learn.

job card **1**	name and fill	51
job card **2**	match, label, draw, write	52
job sheet	plastic vials	53
job sheet	glass jars	54
job card **3**	please fill my cup	55
job sheet	draw how full	56
job card **4**	overfill, underfill	57
job card **5**	solve the puzzle books	58
preparation	which holds more?	59
preparation	which holds less?	60
job card **6**	solve the puzzle books	61
preparation	equal sums?	62
preparation	equal products?	63
job card **7**	solve the puzzle books	64
preparation	what is the sum?	65
preparation	what is the sum or product?	66
job card **8**	on your own (lentil cookies)	67

Basic Materials: Quantities define maximums needed to support any one Job Card in this chapter. Store *high-quantity basics* (Job Boxes, liters of lentils, bottle lids, scoops, funnels) on and under a table or counter. Store *low-quantity basics* near the "basics" sign (see page 49) or in a "basics" box. Consult our Glossary on pages 6-10 for a full description of these items. See the next page for additional special materials used in this chapter.

- [ ] 1 job box
- [ ] 2 liters of lentils
- [ ] 1 scoop
- [ ] 1 funnel
- [ ] story paper

Store these chapter-specific items together in a designated place. They require about 1/2 square foot of dedicated space. General classroom materials (like scissors and tape) are also listed below when used, while others (like pencil and paper) are always assumed.

---

set of 10 containers: 4 glass jars and 6 plastic vials (compare 1, 2, 3, 4, 5, 6, 7, 8)

Tape a symbol label, as detailed below, to each container in this set. Nest the smaller containers inside larger ones. Store them all together in a **gallon storage jug**. Paper labels are provided for 3 identical sets. You will need this many to serve a class of 30 students.

GLASS JARS

• CIRCLE: An **8 fl oz mayonnaise jar** or similar product. We used the smallest Best Foods brand mayonnaise jar available. Other brands may vary slightly in volume, but probably not enough to affect quantitative outcomes.

• TRIANGLE: a **6 oz baby food jar** (BFJ).

• MOON: a **4 oz baby food jar**.

• CLOUD: a **2½ oz baby food jar**. Look for the newer Gerber brand wide-mouth version, redesigned so the same lid fits all sizes. Its capacity is slightly larger than the older small-mouth version.

PLASTIC VIALS

• SQUARE: A **T–60 dram I-O plastic vial**, also called a standard cup. Do *not* substitute I-O 60 dram vials without child-proofing nubs around the mouth. These hold significantly less.

• SUN: A **T-40 dram I-O plastic vial**.

• STAR: A **T-30 dram I-O plastic vial**.

• FISH: A **T-16 dram I-O plastic vial**.

• PERSON: A **T-13 dram I-O plastic vial**. Some pharmacies do not to carry this size because it is so similar to a T-16. Ask at other stores, or purchase from TOPS. Alternately, cut to size a T-16 vial to 65 mm, the height of an equal-diameter T-13.

• HEART: A **T- 8½ dram I-O plastic vial**.

6 puzzle books (compare 5, 6, 7)

Make single photocopies of the 6 line masters. Assemble as directed under booklets. Store these in a **clear cup**.

Masking tape label: *6 puzzle books.*

Sign for Basic Materials

BASICS

baby food jars (small, medium, large)
clear cups
craft sticks
equation tags
extra cups jug
four nesting containers
small bottles (half liter size)
measuring bottle(s)
paper plates
story paper
tubs

Cut and glue to a 4x6 inch index card. Fold it in half to create a stand up sign. Gather the listed materials in small quantities for storage next to this sign. Alternatively, cut this sign in half. Glue both pieces to a grocery bag that has been cut to size, or to a box of appropriate size, and store these materials inside. These *low-quantity basics* support Job Cards in two or more chapters. An "extra cups" label is provided on page 85 for your convenience.

6 puzzle books
(compare 5, 6, 7)

set of 10 containers:
4 glass jars and 6 plastic vials
(compare 1, 2, 3, 4, 5, 6, 7, 8)

C / COMPARE
special materials

COMPARE
4 jars
(circle, triangle, moon, cloud)
6 vials
(square, sun, star, fish, person, heart)

COMPARE
4 jars
(circle, triangle, moon, cloud)
6 vials
(square, sun, star, fish, person, heart)

COMPARE
4 jars
(circle, triangle, moon, cloud)
6 vials
(square, sun, star, fish, person, heart)

CHAPTER SIGN: Glue to a 4 x 6 inch index card, and fold in half. Stand this sign where you store special materials for this chapter. Or cut this sign in half, and glue both pieces to a grocery bag or to a box.

Apply each label to a <u>gallon storage jug</u> with clear packaging tape. (Three duplicate sets recommended for large class sizes.)

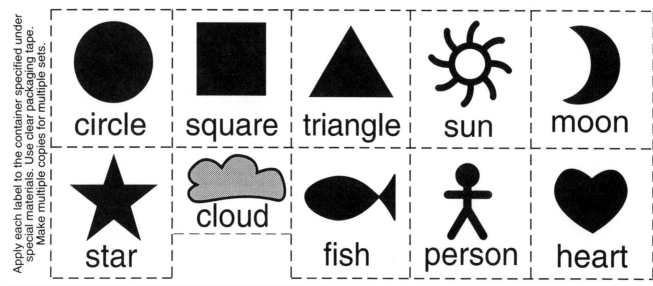

Apply each label to the container specified under special materials. Use clear packaging tape. Make multiple copies for multiple sets.

circle · square · triangle · sun · moon

star · cloud · fish · person · heart

Preparation: 1 copy

TEACHING NOTES

Purpose

To identify each symbol and fill the container fair and full.

Introduction

◗ Hold up each container and ask your class to name the symbol. Repeat as necessary until most can recognize and say each symbol on sight.

◗ Demonstrate how to fill a container *fair and full*. This is always a 2-step process:

1. *Overfill* with lentils, leaving a mound on top.

2. *Shake once* to remove this excess. (The top will remain slightly rounded.)

Fill several more containers fair and full, and stand them in a row. Repeat this 2-step process out loud as you proceed: *overfill* and *shake once*. Invite volunteers to do the same until all 10 containers are lined up, fair and full.

◗ Notice with your class that each fair and full container is slightly rounded on top. You can't scrape your finger across the top without spilling lentils. Try it.

◗ Finally, demonstrate common errors that your class should avoid and you should correct any time you see them:

• Overfill and **pat down**. This crowds extra lentils into the container, leaving it *more* than fair and full.

• Overfill and **scrape level**. This drags away the top layer of lentils, leaving the container *less* than fair and full.

Focus

◆ Which containers are called jars? *The glass ones.* Which ones are called vials? *The plastic ones.*

◆ Can you name (the symbols on) 4 glass jars and fill them fair and full?

◆ Can you name (the symbols on) 6 plastic vials and fill them fair and full?

Checkpoint

◆ How many glass jars are there? Name them.

◆ How many plastic vials are there? Name them.

◆ Show me how to fill the _____ fair and full.

C/1 You need...

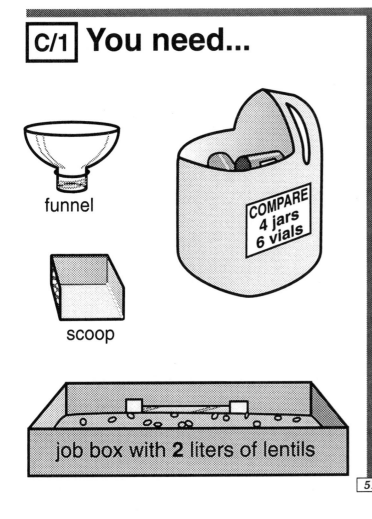

funnel

scoop

job box with **2** liters of lentils

compare 1 ♀ or ♀♀

Name and fill.

Fair and Full.

4 glass jars

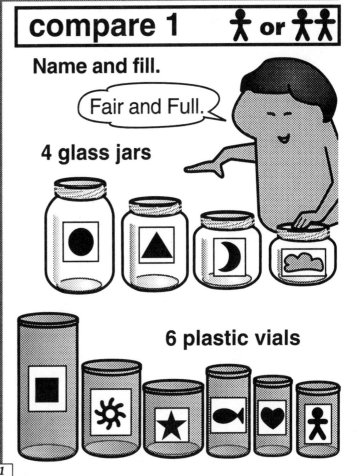

6 plastic vials

LESSON NOTES

Purpose

To write and draw each symbol on its matching container. To order the jars and vials by size.

Introduction

◗ Hold up the Job Sheet showing jar shapes (not vials). Match any jar to its side view on the sheet. Show how to look at the page straight on to see that it fits the outline. Then verify that its top view (a circle) also matches. Is this the tallest / shortest jar?

◗ Ask a volunteer to repeat this exercise, matching one of the plastic vials to its representation on the other Job Sheet. They should first match the side view (a rectangle), then the top view (a circle). Is this the tallest / shortest vial? Is this a wide / narrow vial?

Focus

◆ Which Job Sheet will you do first?

◆ What will you draw on this label? What will you write?

Checkpoint

◆ Let me see your Job Sheets.
  • Which is the tallest jar? The shortest?
  • Which is the tallest vial? The shortest?
  • Which is the widest vial? The narrowest?

◆ Which jars have equal width? *The triangle, moon and cloud have equal width.* Which vials? *The square, sun and star have the same width; the fish and person also have the same width.*

More

◆ Divide a piece of paper into 10 flash cards. Write the symbol names on one side and draw the symbols on the opposite side. Learn to spell all 10 words from memory.

◆ Draw labeled **top views** of all 10 containers. (Hold the top firmly against your paper, then reach way around with your other hand to draw one continuous circle.)

◆ Draw labeled **side views** of all 10 containers. (Trace the outline of each container, lying on its side. Keep the pencil straight up and down.)

C/2 You need...

COMPARE
4 jars
6 vials

2 job sheets

Fill in each label: Draw and write.   compare 2 A

Fill in each label: Draw and write.   compare 2 B

compare 2   🚶 or 🚶🚶

Match each shape.

Fill in each label: Draw *and* write.

Draw and write on each label.

sun

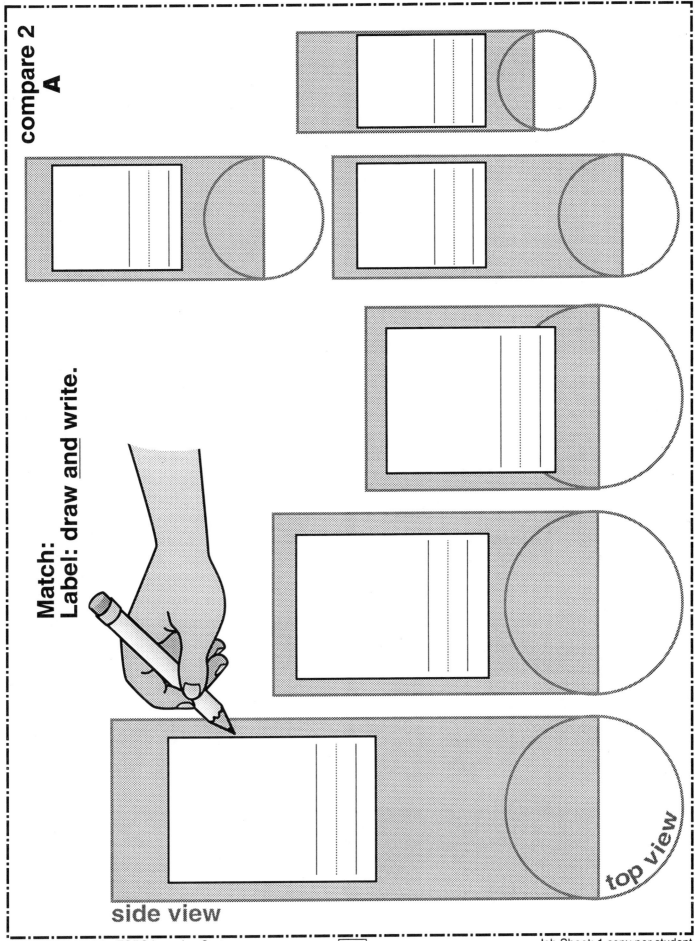

compare 2
A

Match:
Label: draw <u>and</u> write.

side view

top view

compare 2
B

side view

top view

Match:
Label: draw <u>and</u> write.

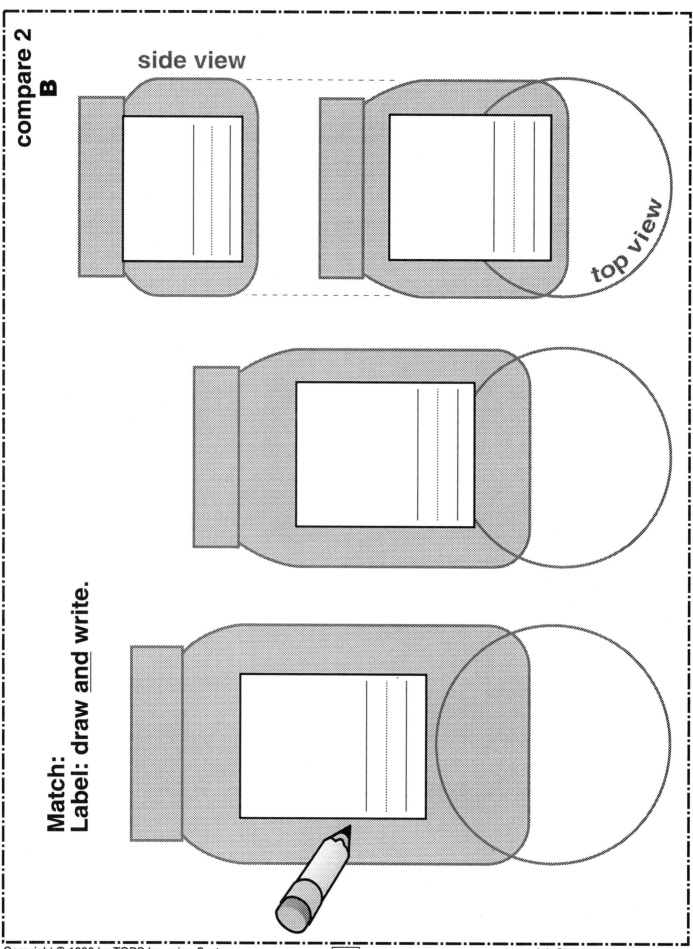

 Job Sheet: 1 copy per student

TEACHING NOTES

Purpose

To distinguish between 4 degrees of fullness.

Introduction

◗ Fill one of the larger containers to each degree of fullness:

• **HEAPING Full:** Overfill. Don't shake away the mound on top.

• **FAIR and FULL:** Overfill, then shake once. (Don't pat down. Notice that the top remains *slightly* rounded.)

• **LEVEL:** Overfill, then scrape level with your finger. Notice that the top layer of lentils gets scraped out of the container.

• **HALF Full:** Fill until the empty space *above* equals the filled space *below*. (Compare spaces between thumb and forefinger.)

◗ Draw examples of each of degree fullness on your blackboard. Model how to represent the lentils with shading and an occasional circle. Define the top surface with a solid line.

◗ Compare your blackboard drawing of FAIR and FULL to your drawing of LEVEL. Children will miss the subtle difference between these drawings unless you point it out: "Fair and Full" is slightly higher than the top while "Level" is slightly lower.

Focus

◆ Let's play "Store:"

• Who will be the storekeeper first, and who will be the customer?

• What container will you use?

• How full will you make it?

Checkpoint

◆ Name 4 different degrees of fullness.

◆ Show me the _____ container. Fill it _____.

◆ Show me your drawing. Did you need to draw every lentil? How did you show the top of the lentils?

◆ Is a Fair and Full cup perfectly flat on top? *No, it is slightly rounded.*

◆ If a Heaping Full cup is patted down, does this make it Fair and Full? *No. Extra lentils get crammed into the cup, making it* more *than fair and full.*

◆ If a Heaping Full cup is scraped level, does this make it Fair and Full? *No. The top layer of lentils gets scraped out of the container, making it* less *than Fair and Full.*

C/3 You need...

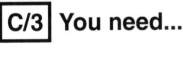

job sheet

job box with **2 liters of lentils**

Draw

Draw how full.

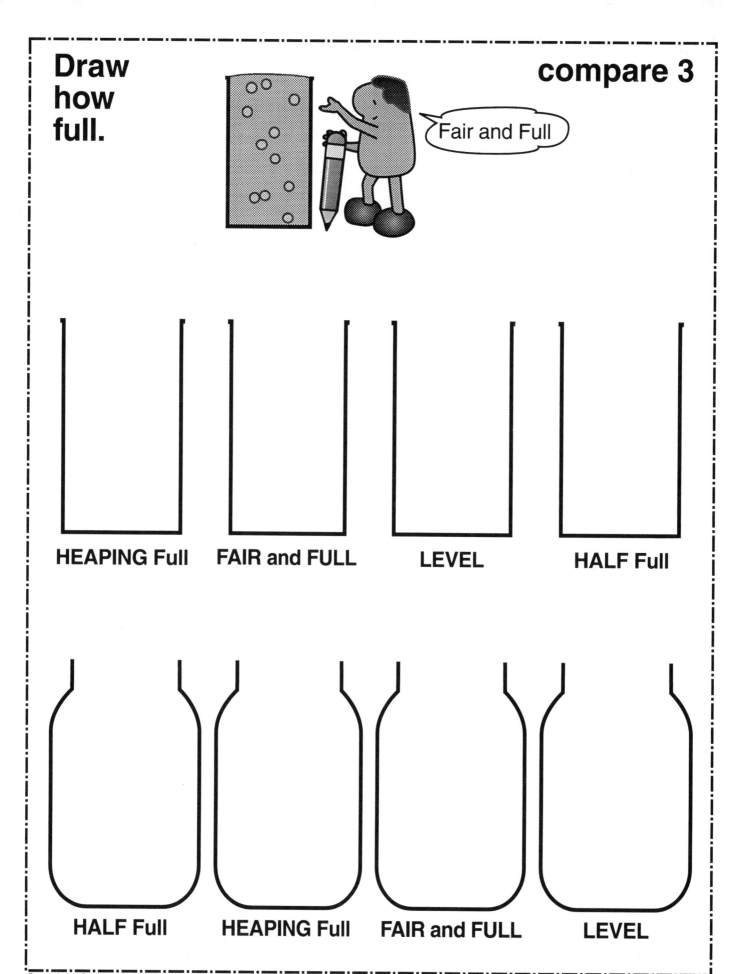

compare 3

Fair and Full

HEAPING Full   FAIR and FULL   LEVEL   HALF Full

HALF Full   HEAPING Full   FAIR and FULL   LEVEL

Job Sheet: 1 copy per student

LESSON NOTES

Purpose

To decide by experiment and observation which container holds more and which holds less.

Introduction

◗ Set the square and the triangle side by side. Ask which container holds less. Everyone will recognize that the triangle holds less because it looks smaller.

◗ Now ask a volunteer to *prove* this by pouring lentils. (Children are likely to fill both containers fair and full, then argue that one *looks* bigger or *feels* heavier than the other. Ask them to empty either container, then pour the full into the empty.)

◗ Write these important observations on your blackboard:

Large overfills small.
Small underfills large.

Focus

◆ Choose any 2 containers. Can you pour lentils to decide which holds more and which holds less? (Fill only one of the containers fair and full, then pour it into the second.)

◆ Can you write down what you just observed? (Example: A underfills B, so A is smaller and B is larger.)

Checkpoint

◆ Let me see what you have written. Show me that this sentence is really true.

◆ Here are 2 jars. Show me by experiment which is smaller.

More

◆ Line up the Glass/Plastic/All containers from largest to smallest and list your results. Prove that your lineup is correct. (Fill the largest container and pour it into the next largest, then pour that full container into the next largest, all they way down the line. If the lineup is correct, each container will overfill its next smaller neighbor.)

◆ Draw a picture of a large container overfilling a small container. Then draw the opposite.

C/4 You need...

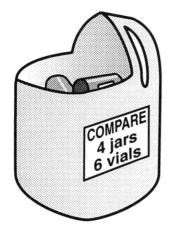

COMPARE
4 jars
6 vials

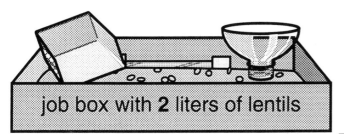

job box with **2 liters of lentils**

---

compare 4 🧍 or 🧍🧍

Pour. Look. Write.

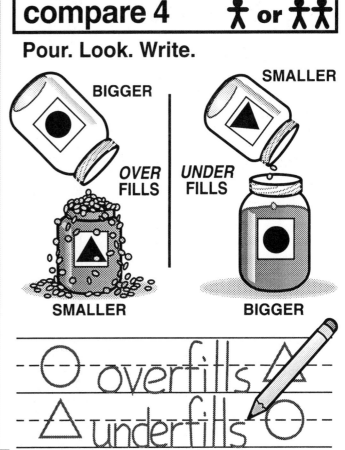

BIGGER   SMALLER

OVER FILLS   UNDER FILLS

SMALLER   BIGGER

◯ overfills △
△ underfills ◯

LESSON NOTES

Purpose

To decide by experiment which container holds more and which holds less.

Introduction

◗ Hold up each Puzzle Book. Discuss a general strategy for solving the 8 problems inside.
- Read the question.
- Pour lentils to decide.
- Turn the page to see if you are right.

◗ Tips on using the books:
- Each leaf is folded into a double layer. Don't try to separate them.
- Open the pages all the way and crease them flat near the spine. Hold the book open by pushing its edges into the lentils.

Focus

◆ Which book will you try first?

◆ How will you solve each puzzle? *Read. Pour. Confirm by turning the page.*

◆ Can you solve this puzzle?

Checkpoint

◆ Which Puzzle Books did you try?

◆ Did you solve all 8 puzzles? Show me how to solve this one.

C/5	**You need...**

compare 5 A · holds **more?** ▲/●

compare 5 B · holds **less?** ☾/✖

puzzle books

COMPARE 4 jars 6 vials

job box with **2** liters of lentils

---

compare 5	🧍

Experiment. Then turn the page.

compare 5 A

holds more? ▲/●

compare 5 B

holds less? ☾/✖

Can you invent your own puzzles?

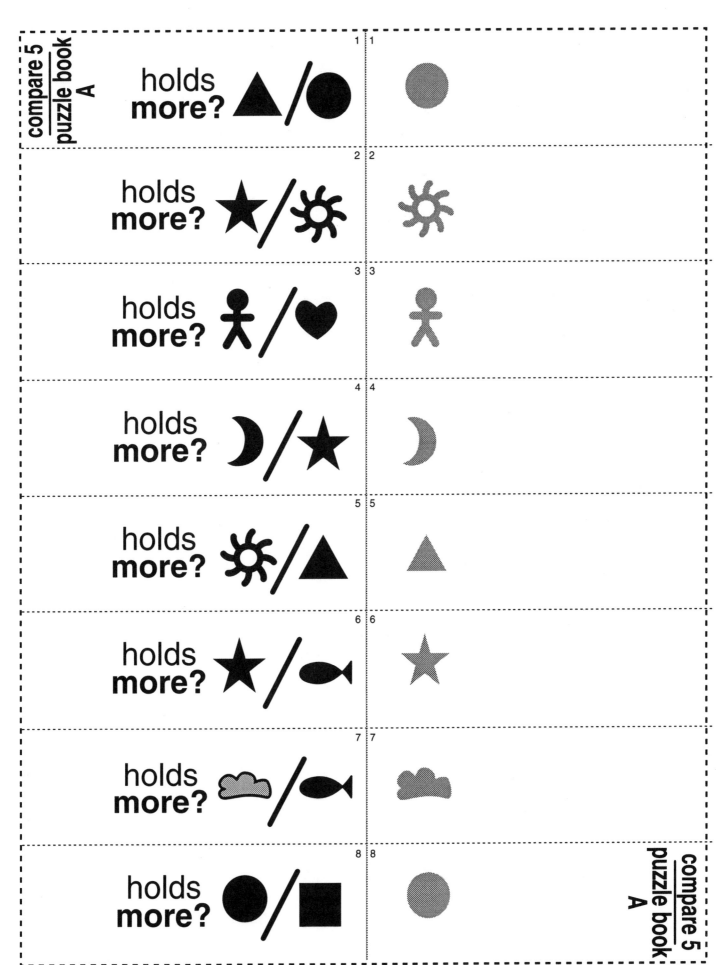

1 | 1 — holds **more?** ▲/●

2 | 2 — holds **more?** ★/☀

3 | 3 — holds **more?** 🧍/♥

4 | 4 — holds **more?** 🌙/★

5 | 5 — holds **more?** ☀/▲

6 | 6 — holds **more?** ★/🐟

7 | 7 — holds **more?** ☁/🐟

8 | 8 — holds **more?** ●/■

Preparation: 1 copy

1 | 1

holds
less?

2 | 2

holds
less?

3 | 3

holds
less?

4 | 4

holds
less?

5 | 5

holds
less?

6 | 6

holds
less?

7 | 7

holds
less?

8 | 8

holds
less?

Preparation: 1 copy

LESSON NOTES

Purpose

To decide by experiment if several smaller containers equal a larger one. To solve equations.

Introduction

▶ Equations say what is equal. What we write on one side of the equal sign is the *same amount* as what we write on the other side:

    2 dozen eggs = 24 eggs
    60 minutes = 1 hour
    1 dollar = 2 quarters + 5 dimes

▶ Equations can also make puzzles with missing pieces to fill in. Write examples like these and ask your class for answers:

    ___ pennies = 1 nickel
    this class = ? students
    4 days + x days = 1 week
    this pie ⊗ = ___ pieces

▶ Hold up both Puzzle Books. Solve the front page of 6A to model a problem-solving strategy:

• Locate all needed containers, in this case, the triangle, heart and square.

• Fill the containers on the left side of the equation fair and full. See if these fill the container on the right side of the equation fair and full.

• Turn the page to confirm your result.

Focus

◆ Which book will you try first?

◆ How will you solve each puzzle? *Read. Pour. Check the answer on the next page.*

◆ Can you solve this puzzle?

Checkpoint

◆ Which Puzzle Books did you try?

◆ Did you solve all 8 puzzles? Show me how to solve this one.

Special Note

The equalities in these Puzzle Books are always very close, and the inequalities are always far enough apart to be obviously unequal. Yet some children may hold an unrealistically high standard, so that very close, for them, is not close enough.

Ask these children to first repour the equation, to demonstrate good technique. If they still insist that close is not equal, compromise with "almost" equal. Scientific literacy includes recognizing that real-world measurements always contain a degree of unavoidable error. With experience, perfectionists will learn to discriminate between acceptable and unacceptable limits of error.

	**1** **1** Yes. Equal!
	**2** **2** No. Less than ▲
	**3** **3** Yes. Equal!
	**4** **4** No. More than
	5 **5** No. More than ☀
	**6** **6** Yes. Equal!
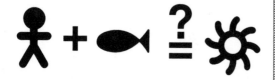	**7** **7** Yes. Equal!
	8 **8** No. Less than ☀

Preparation: 1 copy

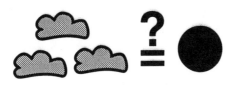

1 1 Yes. Equal!

2 2 No. Less than

3 3 Yes. Equal!

4 4 No. More than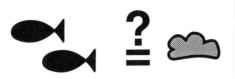

5 5 Yes. Equal!

6 6 No. Less than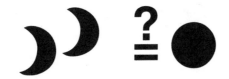

7 7 No. More than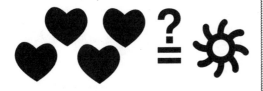

8 8 Yes. Equal!

Preparation: 1 copy

Purpose

To search by trial and error for one particular container that holds as much as a group of containers.

Introduction

Hold up both Puzzle Books. Solve the front page of 7A to model a problem-solving strategy:

• First locate all containers named in the problem, in this case, the square and the heart. Fill these fair and full.

• Now choose another container that appears large enough to hold these smaller portions. Test by trial and error until you find the right one.

• Turn the page to confirm your result.

Focus

◆ Which book will you try first?

◆ How will you solve each puzzle? *Read. Pour. Check the answer on the next page.*

◆ Can you solve this puzzle?

Checkpoint

◆ Which Puzzle Books did you try?

◆ Did you solve all 8 puzzles? Show me how to solve this one.

C/7 **You need...**

compare 7
A
■ + ♥ = ?

compare 7
B
🧍🧍 + 🌙 =

puzzle books

COMPARE
4 jars
6 vials

job box with **2** liters of lentils

compare 7

Try these Puzzle Books:

compare 7
A
■ + ♥ = ?

compare 7
B
🧍🧍 + 🌙 = ?

Can you write other equations to solve?

1 | 1

■ + ♥ = ?

2 | 2

🐟 + 🧍 = ?

3 | 3

☀ + ♥ = ?

4 | 4

☁ + 🌙 = ?

5 | 5

♥ + 🐟 = ?

6 | 6

 = ?

7 | 7

🐟 + ☁ + ★ = ?

8 | 8

♥ + 🐟 + 🧍 + ☁ = ?

Preparation: 1 copy

1 | 1

2 | 2

3 | 3

4 | 4

5 | 5

6 | 6

7 | 7

8 | 8

Preparation: 1 copy

LESSON NOTES

Purpose
To provide students with an approved way to pursue their own ideas about comparing volumes.

Introduction
Display this card when you want to do the suggested activity or experiments that you design yourself.

Focus
◆ What do you want to study about comparing?
◆ Are these the materials you need?

Checkpoint
Report in words and pictures:
• What you did.
• What you learned.
• Questions you may still have.

Special Note
Any great ideas? Record them on the opposite page.

C/8 You need...

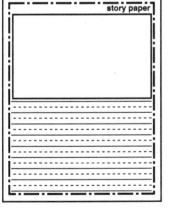

story paper

COMPARE
4 jars
6 vials

job box with **2 liters of lentils**

 Ask your teacher for other items.
Tell why you need them.

---

compare 8   👤 or 👤👤

On your own.

Cookie Recipe   story paper

2 ● flour
3 ★ sugar
1 🐟 butter
3 ▣ lentils
1 egg

Yum. Lentil cookies.

D / DESIGN

In this chapter: What are lentils good for? They are good for writing letters and numbers; for drawing patterns and shapes; for sculpting lakes, rivers and oceans; for burying treasures and drawing maps; for building roads and directing people to go north or south or east or west or right or left. Lentils are good for pretending to be a doctor, chemist, baker or magician, or even a hockey player!

job card **1**	write and save	71
job card **2**	draw it small, medium, and large	72
preparation	sixteen shapes	73
job sheet	drawing paper	74
job card **3**	create a landscape	75
preparation	landforms book	76
preparation	landscape / townscape props	77
job card **4**	hide treasure and draw a map	78
job card **5**:	design a town	79
preparation	street sticks / more landscape	80
preparation	street sticks / more townscape	81
job card **6**:	welcome to Planet Lentil!	82
job card **7**:	on your own (lentil hockey)	83

Basic Materials: Quantities define maximums needed to support any one Job Card in this chapter. Store *high-quantity basics* (Job Boxes, liters of lentils, bottle lids, scoops, funnels) on and under a table or counter. Store *low-quantity basics* near the "basics" sign (see page 49) or in a "basics" box. Consult our Glossary on pages 6-10 for a full description of these items. See the next page for additional special materials used in this chapter.

- [ ] **1 job box**
- [ ] **up to 2 liter of lentils**
- [ ] **story paper**
- [ ] **baby food jars: small, medium, large**
- [ ] **1 scoop**
- [ ] **1 funnel**
- [ ] **2 craft sticks**

☞ *Please observe our copyright restrictions on page 2.*

Store these chapter-specific items together in a designated place. They require about 1/4 square foot of dedicated space. General classroom materials (like scissors and tape) are also listed below when used, while others (like pencil and paper) are always assumed.

spoon (design 1, 2, 3, 4, 6, 7)

Metal feels better than plastic when scraped across the bottom of the cardboard box. Teaspoon size works best.

canning ring (design 1, 2, 3, 4, 6, 7)

Use the smaller standard size, if available.

book of shapes (design 2)

Photocopy and assemble the booklet.

landforms book (design 3)

Photocopy and assemble the booklet.

landscape props (design 3, 4, 6, 7)

1. Make one photocopy each of the landscape/ townscape line masters. (Find these on pages 77, 80 and 81. Set the street sticks labels aside.)

2. Cut each strip of props along the zigzagged edges. Cut out the 3 hexagons at the end of one landscape strip, and press each one into a bottle cap.

 4. Crease the remaining strips along the 3 long dotted lines before you separate them. Fold back on the center line so graphics show on both sides, then fold up the pointed grey tabs.

5. Cut the strips apart on the solid lines to make separate stand-up figures, and press their grey bases into bottle caps. No glue is necessary.

6. Store the landscape props in a clear cup. Label a clothespin "N" on both wings and add this to the cup.

Paper label provided.

townscape props (design 5, 6, 7)

Store the townscape props in another clear cup.
Paper label provided.

street sticks (design 5)

You have already copied these labels for 12 avenues and 12 streets with the landscape props. Cut out each double name (24 rectangles in all) and glue them to 24 **craft sticks**:

1. Rub or brush a thin film of white **glue**, or use a glue stick, on both sides of a craft stick, stopping just short of each rounded end. You might want to work over paper to keep your table clean.

2. Holding the stick by an end, press a narrow sticky edge along the center of a street label lying face down on your table.

4. Pick up the stick with the street label sticking to it, and turn it over. The grey line on a flat paper "roof" is now centered over a slightly longer wood "wall."

5. Rest the bottom edge of this wall on your table and press the roof down onto both sticky sides. Only each end of the craft stick and its bottom edge remain uncovered.

When the glue is dry, bundle the sticks in a rubber band.

street sticks **(design 5)**

townscape props (design 5, 6, 7)

landscape props (design 3, 4, 6, 7)

landforms book (design 3)

book of shapes (design 2)

canning ring (design 1, 2, 3, 4, 6, 7)

spoon (design 1, 2, 3, 4, 6, 7)

LANDSCAPE
props
bottle cap figures
craft stick

TOWNSCAPE
props
bottle cap figures

Apply each sign to a clear cup with
clear packaging tape.

D / DESIGN
special materials

CHAPTER SIGN: Glue to a 4 x 6 inch index card, and fold in half.
Stand this sign in the space where you store special materials for
this chapter. Or cut this sign in half, and glue both pieces to a
grocery bag that has been cut to size, or to a box.

Preparation: 1 copy

TEACHING NOTES

Purpose

To practice writing letters and numbers.

Introduction

▶ Spread a liter of lentils across the bottom of a Job Box. Give it a few quick shakes to evenly cover the bottom. Scrape out a simple shape like the number "1" with a canning ring. Record what you wrote on your blackboard.

▶ "Erase" the lentils with another shake. Scrape out something a little more complicated, like the letter "A". Notice how lentils scraped from new lines fall into old lines where they meet or cross. Remove these stray lentils with the spoon, then retrace until the letter is well formed. Record this letter on your blackboard.

▶ Give the box another shake to smooth the lentils. Invite volunteers to write their own letters and numbers, following this computer-like procedure: WRITE in the lentils; SAVE to the blackboard; ERASE with a quick shake of the box.

Focus

◆ Will you write letters or numbers in the lentils?

◆ Where will you save your work? *On writing paper.*

◆ How will you erase what you wrote? *With a quick shake of the box.*

Checkpoint

◆ Show me how to write the letter ____; the number ____.

◆ May I see your written work?

More

▶ Start with an empty Job Box and a full liter of lentils. Write letters and numbers by pouring out lentils in a small ribbon, as if decorating a cake.

▶ How is this kind of lentil writing different than what you did before? *Now, lentils (positive shapes) form the letters. Before, the floor of the box (negative space) formed the letters.*

▶ What is the very smallest letter you can "write" with lentils? How many lentils do you need? *The letter "i" can be formed with two lentils dotted by a third, or "l" with 3 lentils in a row. Very clever students might use a single lentil; on edge for "l" or lying flat for "o."*

D/1	**You need...**

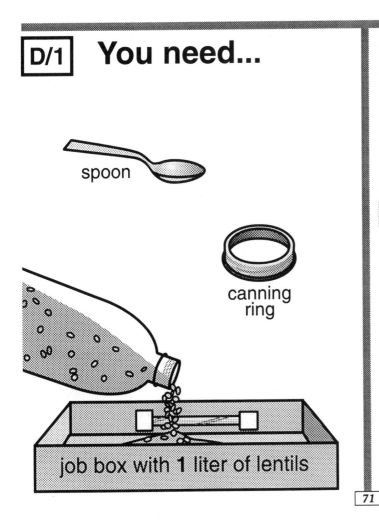

spoon

canning ring

job box with **1 liter of lentils**

design 1	🚶

Write!

Save.

Aa   Bb   Cc
1   2   3   4

◀ **Erase** ▶

TEACHING NOTES

Purpose

To draw geometric shapes scaled to small, medium and large sizes.

Introduction

▶ Draw 3 circles of increasing size side by side on your blackboard. Label them small, medium and large. Point out that these circles have different sizes, but the same shape. (They are proportional.)

▶ Erase the circles but not the labels. Draw 3 squares of increasing size next to the correct labels. Ask how these drawings are the same, and how they are different. *They all have the same shape; they all have different sizes.* (They are proportional.)

Focus

◆ Look at the Book of Shapes: Which of these small shapes will you draw first?

◆ What tools will you use to make a large-sized (and proportional) copy of this shape in the lentils? *The canning ring and the spoon.*

◆ Where will you draw a medium-sized copy of this shape? *On the Drawing Paper.*

Checkpoint

◆ Let me see your drawings. How is this drawing the same as the one in the Book of Shapes? How is it different? *Same shape; different size.* (Both drawings are proportional.)

More

◆ Clear away lentils to make an open space in the middle of your Job Box, then gently shake the box to close this space back up again. Stop when you recognize a new shape forming. Name it. *Cloud, noodle, shoe, sleeping cat, etc.* Draw it.

◆ Repeat the same shape but with different proportions:

• What happens when you increase the length of a square, but not its width? (You get a rectangle.)

• What happens when you shrink a circle in one direction but not the other? (You get an oval.)

• What other shapes can you change?

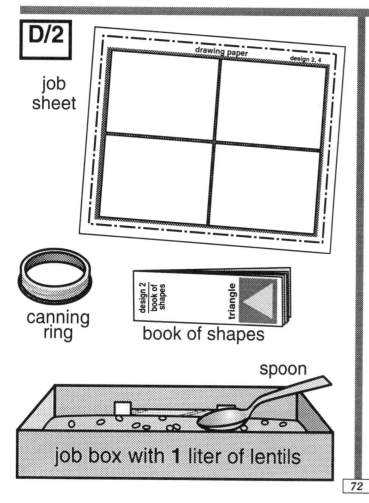

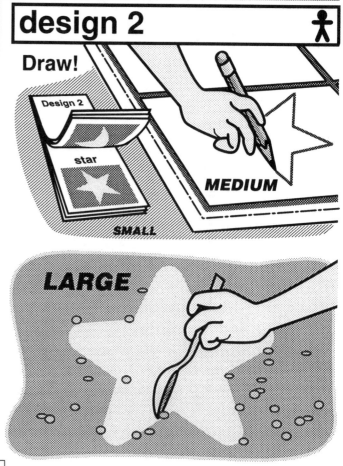

triangle 1	diamond 2
square 3	circle 4
heart 5	moon 6
star 7	sun 8
egg 9	ring 10
person 11	pentagon 12
fish 13	rhombus 14
cloud 15	trapezoid 16

Preparation: single copy

design 2, 4

drawing paper

Job Sheet: several copies per student

TEACHING NOTES

Purpose

To model, draw and write about a wilderness landscape. To exercise creativity and imagination.

Introduction

◗ Imagine that you are taking a vacation in a wild place, far away from people. Look through the Landforms Book for inspiration and ideas. Decide as a class where you would like to visit. Then "go there" by arranging lentils in a job box.

◗ Carve a river / lake / beach into the lentil "land mass." Then ask what other kinds of things you might see, or smell, or hear, or touch, or taste in this environment. Brainstorm a list of natural objects.

◗ Add bottle-cap Landscape Props to your lentil wilderness. Refer to your brainstormed list, and draw an original prop on folded scratch paper. Stand this "tent" style, without a bottle cap base, among the lentils.

◗ Improvise a class story about the special place you have created. Begin with an introductory sentence like, "Once upon a time our class went on a summer vacation…." Or perhaps the classic, "It was a dark and stormy night…." Ask volunteers to each contribute an additional sentence that builds a story chain. Call forth the naturalist in your students by prompting them to describe in rich detail the worlds they see in their imaginations.

Focus

◆ What kind of landforms will you create? *Creek, river, delta, small lakes, large lake, interconnected lakes, ocean beach, ocean island, bay, penninsula, straits and narrows, ocean islands, lake islands, mountain range, volcano,…*

◆ Which Landscape Props will you use? (Not too many: you'll need to draw each one you use.)

◆ Will you include your own natural props? (Cut and fold paper rectangles to draw on. Stand these A-frame style, in the lentils. Children may tape or glue these into their final drawings.)

Checkpoint

◆ Show me your drawing and story. What can you tell me about it?

◆ Did you draw any original props?

D/3 You need...

story paper

landforms book

job box with **2 liters** of lentils

design 3 � or ��

Create a landscape.

design 3
landforms
river

Draw it.
Write a story.

story paper

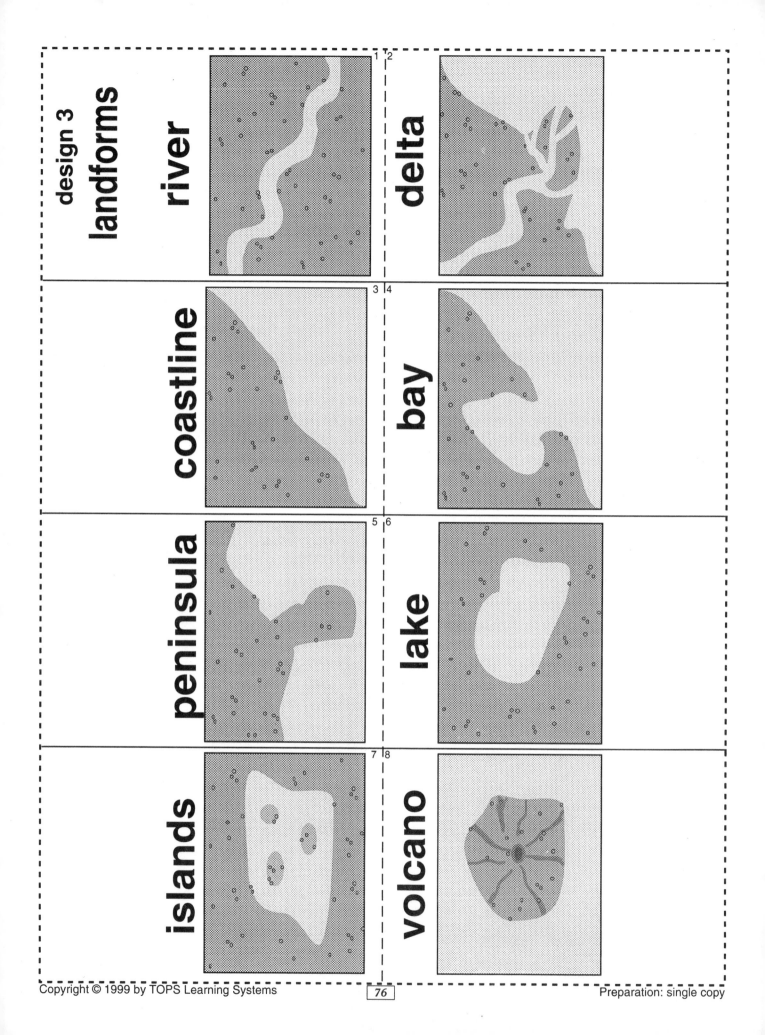

design 3
landforms

river

delta

coastline

bay

peninsula

lake

islands

volcano

Preparation: single copy

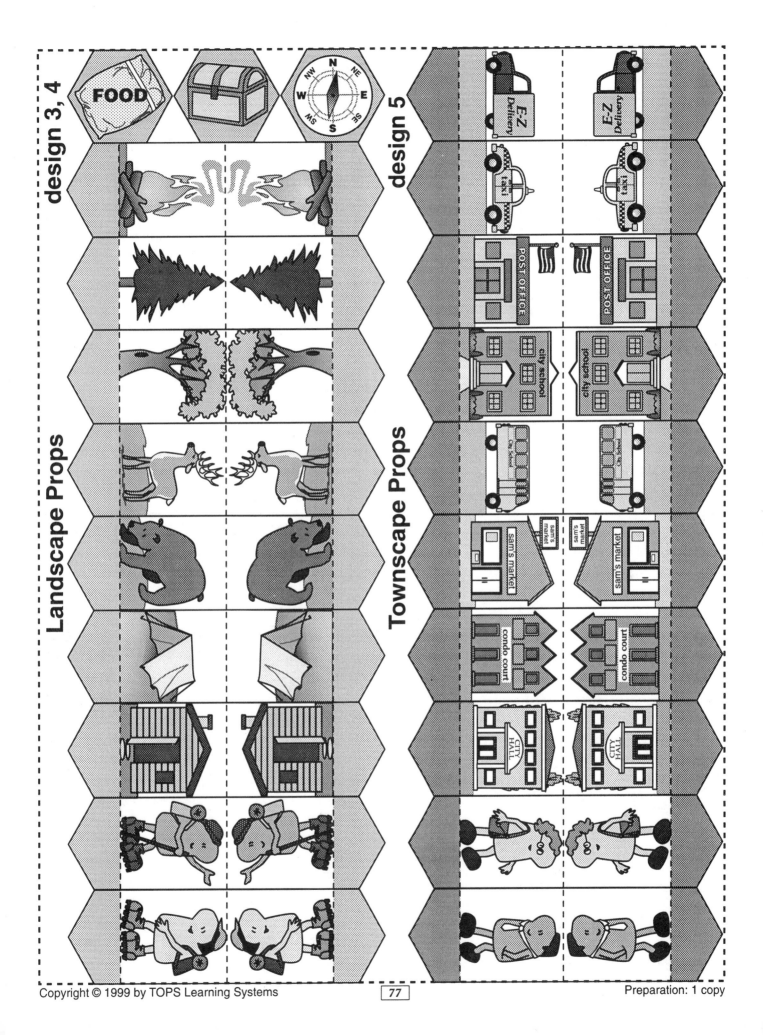

design 3, 4

Landscape Props

FOOD

design 5

Townscape Props

Preparation: 1 copy

TEACHING NOTES

Purpose

To hide a "treasure" within a landscape. To accurately map its location so someone else can easily find it.

Introduction

◆ Create a country landscape. Then bury the Landscape Prop "treasure chest" anywhere in the box, near a "tree" or other landmark, perhaps. Call attention to the bold letter "N" at the top of the Job Card. This defines the far side of the box as "north" when you put the Job Card in its holder.

◆ Draw a blackboard map that represents the landscape you have just modeled. Write "N" at the top of your drawing to correspond to the back of the Job Box. Sketch in only major features. Don't draw each lentil.

◆ Decide with your class where you will place the "X" on your map. You want to be accurate, so anyone who reads the map can directly locate the treasure. To promote a lively discussion, mark the X where the treasure is not, then allow students to talk you out of your "error" by arguing where the X really belongs.

◆ Erase this mark so you can bury the treasure in a new location. Choose a volunteer to leave the room before you do this. Write a new X on your map to mark the new spot, then invite the volunteer back in to read your map and locate the treasure. If she finds it immediately, without tearing up the landscape, congratulate both the map reader and the map maker. Emphasize the win-win nature of this process. (Children tend to feel they somehow win when the map reader loses, and can't find the treasure. On the contrary! A quick find is a sign of the mapmaker's success.)

Focus

◆ What kind of a landscape will you create? What objects will you include? (Two students might cooperatively create a landscape, then map it individually. They might take turns at hiding and seeking by burying the treasure in different places.)

◆ How will you know if you made a good map? *The treasure will be easy to find.*

Checkpoint

◆ Show me your treasure map(s). Did you draw important features, and leave out what was not important?

◆ Was this map accurate? Why do you think so?

More

◆ Design and hide your own mini-treasures.

◆ Give written directions to the treasure.

D/4 | You need...

job sheet

drawing paper   design 2, 4

LANDSCAPE props

job box with **2 liters of lentils**

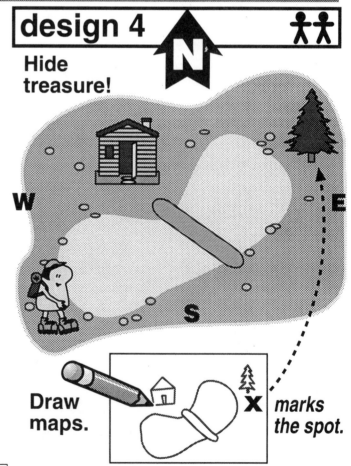

design 4

Hide treasure!

W · E · S · N

Draw maps.

X *marks the spot.*

TEACHING NOTES

Purpose

To design and map a town. To develop a sense of direction in terms of cardinal points and street names. To distinguish right from left.

Introduction

▶ Place this Job Card in a Job Box of lentils so your students can see "N." Lay out the grid of 4 avenues and 4 streets. Should we place these in any logical order? (Arrange avenues in numerical order, named streets in alphabetical order.)

▶ Put the post office prop in one corner of town; place Sam's market, both city dwellers and the car prop in the diagonally opposite corner of town. The woman is giving directions to the man about how to drive to the post office. What should she tell him? (Use N, S, E and W in your description.) Is there more than one way to get to the post office from Sam's market?

▶ Stand behind your students at the back of the room, so you and your class all face forward. Play "Simon Says" to teach right and left: raise your _____ hand; jump on your_____ foot; touch your _____ ear; cover your _____ eye; turn _____.

▶ Repeat the directions from Sam's market to the post office in terms of right and left. (Children who attempt this should be standing next to Sam's market and facing the post office!)

Focus

◆ Can you sort the Street Sticks by name and number?

◆ Will you lay out the streets in any kind of order?

◆ What props will you use?

◆ Can you draw a map and give directions?

Checkpoint

◆ Please tell this visitor how to get from _____ to _____.

◆ Show me your map. Let's see if I can follow your directions.

More

◆ Develop a larger town layout on butcher paper. Write a guided tour of important sites.

◆ Go outside and march toward each of the 4 cardinal points. A great time to do this is at high noon (12:00 standard time, 1:00 daylight savings) when shadows point north (in northern latitudes).

When you are facing *north*…
  south is always *behind* you,
  east is always to your *right*,
  west is always to your *left*.

D/5 You need...

story paper

street sticks

job box with **1** liter of lentils

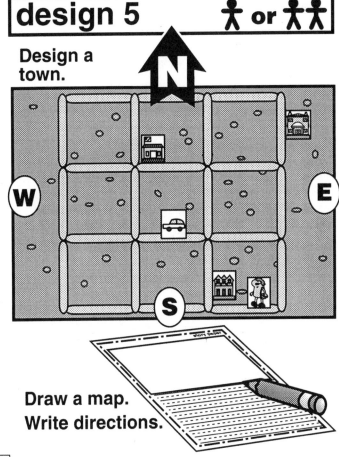

design 5 🚶 or 🚶🚶

Design a town.

Draw a map.
Write directions.

Street Sticks

1. Cut on narrow dashed lines.
2. Fold over craft stick on grey bars and glue.

design 5

1st Avenue	3rd Avenue
1st Avenue	3rd Avenue
1st Avenue	3rd Avenue
1st Avenue	3rd Avenue
1st Avenue	3rd Avenue
2nd Avenue	4th Avenue
2nd Avenue	4th Avenue
2nd Avenue	4th Avenue
2nd Avenue	4th Avenue
2nd Avenue	4th Avenue

More Landscape Props

design 3, 4

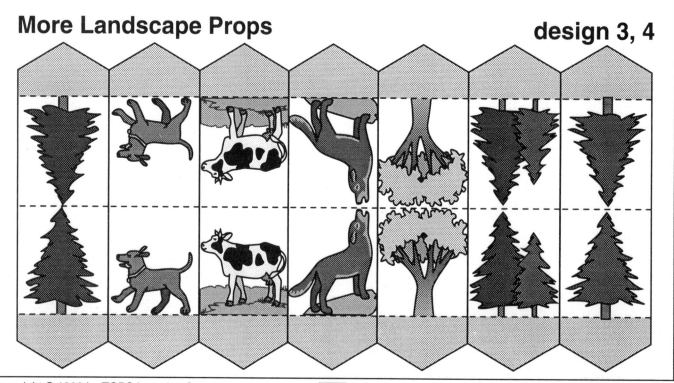

Preparation: 1 copy

Street Sticks

1. Cut on narrow dashed lines.
2. Fold over craft stick on grey bars and glue.

design 5

Aspen Street	Cedar Street
Aspen Street	Cedar Street
Aspen Street	Cedar Street
Aspen Street	Cedar Street
Aspen Street	Cedar Street
Aspen Street	Cedar Street
Birch Street	Dogwood Street
Birch Street	Dogwood Street
Birch Street	Dogwood Street
Birch Street	Dogwood Street
Birch Street	Dogwood Street
Birch Street	Dogwood Street

More Townscape Props

design 5

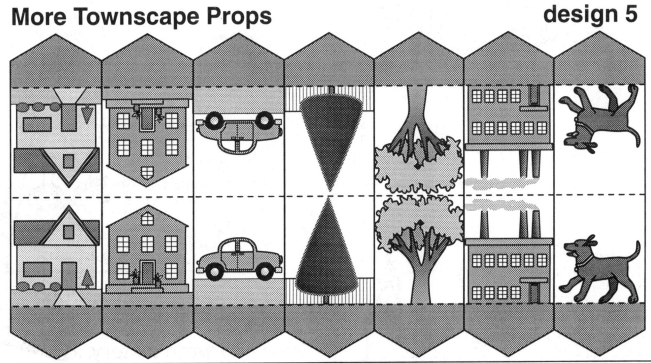

Preparation: 1 copy

TEACHING NOTES

Purpose

To try on different job roles and see how they fit. To exercise the imagination.

Introduction

What will you be when your grow up? Brainstorm a list of possibilities with your class.

Tinker, tailor, soldier, sailor, rich man, poor man, beggar man, thief, doctor, lawyer, merchant, chief, are occupations that spring immediately to my mind. My brother and I recited this childhood rhyme as we slid cherry pits across our dinner plates. When we finished the list with a dozen cherries, we'd go around again, always in the same circle. We had "destiny dessert" about once a month, but never really expanded our horizons.

The idea here is to lead your students beyond closed loops, to imagine themselves in professions they have never thought about before.

How would you like to be a banker, cook, chemist, farmer, pilot, teacher, fireman, librarian, cowboy, engineer, diplomat, peacemaker, hockey player, politician, disc jockey, pharmacist, social worker, printer, actor, psychologist, homemaker, miner, programmer, animal trainer, philosopher, store clerk, artist, preacher, magician, astronaut, police officer, photographer, reporter, web master, curriculum developer…?

Focus

◆ What will you pretend to be?

◆ What will you imagine the lentils to be? *Shannon, age 6, decided to be a banker and said, "Lentils are more fun than a pile of gold coins....Well..., Almost."*

Checkpoint

◆ Tell me what you imagined.

◆ Show me what you drew and wrote.

Purpose

To provide students with an approved way to pursue their own ideas about design. To encourage creativity.

Introduction

Display this card when you want to do the activity shown here, or your own experiments.

Focus

◆ What do you want to study about design?

◆ Are these the materials you need?

Checkpoint

Report in words and pictures:
- What you did.
- What you learned.
- Questions you may still have.

Special Note

Hooray for originality! Record some of your students' most creative ideas on the opposite page.

| D/7 | You may use... |

craft sticks

TOWNSCAPE props

LANDSCAPE props

story paper

story paper

job box

 Ask your teacher for other items. Tell why you need them.

design 7   ♀ or ♀♀

On your own.

My idea:
Lentil Hockey

E / MEASURE

In this chapter: Measure volume by the cupful, or read calibrations on the side of a bottle. Notice that the same one-cup volume can fill containers of different shapes fair and full. Understand that container width matters just as much as container height. Arrange, write and record equations that relate cups to pints to quarts and more.

job card **1**	pour cups	87
job sheet	how high do cups reach?	88
job card **2**	funny cups	89
job card **3**	count whole cups	90
preparation	what do these fill?	91
job card **4**	arrange cup equalities	92
preparation	equation tags	93
job card **5**	count half cups	94
preparation	how many to fill?	95
job card **6**	arrange half cup equalities	96
job card **7**	draw and pour	97
job sheet	how high do half cups reach?	98
job card **8**	on your own (a handful)	99

Basic Materials: Quantities define maximums needed to support any one Job Card in this chapter. Store *high-quantity basics* (Job Boxes, liters of lentils, bottle lids, scoops, funnels) on and under a table or counter. Store *low-quantity basics* near the "basics" sign (see page 49) or in a "basics" box. Consult our Glossary on pages 6-10 for a full description of these items. See the next page for additional special materials used in this chapter.

- [ ] **1 job box**
- [ ] **up to 2 liters of lentils**
- [ ] **1 scoop**
- [ ] **1 funnel**
- [ ] **extra cups**
- [ ] **equation tags**
- [ ] **1 measuring bottle**

Special Materials: Chapter E / Measure

Store these chapter-specific items together in a designated place. They require about 1/2 square foot of dedicated space. General classroom materials (like scissors and tape) are also listed below when used, while others (like pencil and paper) are always assumed.

4 funny cups (measure 2)

Prepare 4 cups of different heights but equal volume. Nest them one inside the other for storage:

TALL CUP: Stick 2 masking tape labels on a dedicated **standard cup**, one above the other.

Top label: *funny cups.*
Lower label: *tall cup.*

MEDIUM CUP: <u>Cut</u> an **8 ounce yogurt tub** <u>to size</u> so it holds 1 standard cup, fair and full.

Masking tape label: *medium cup.*

SHORT CUP: Use a **6 ounce tuna fish can**. Masking tape label: *short cup.*

VERY SHORT CUP: Cut a **quart yogurt tub** or margarine tub to size so it holds 1 standard cup when filled fair and full.

Masking tape label: *very short cup.*

2 puzzle books (measure 3, 5)

Make single photocopies of both line masters. Assemble as directed under <u>booklets</u>.

set of 5 measuring cups (measure 3, 4, 5, 6, 7, 8)

Nest these 5 containers inside each other and store them in a **gallon storage jug**. Extra paper labels are provided. You will need up to 3 sets to serve many students.

HALF CUP: Use a **30 dram plastic vial** as is. (For children capable of precise work, you may want to <u>cut</u> a 40 or 60 dram vial <u>to size</u> as on page 103.)

Masking tape label: *half cup.*

CUP: Use a **60 dram standard cup**.

Masking tape label: *cup.*

PINT: Cut a **half-liter bottle** to size so it holds 2 standard cups when filled fair and full.

Masking tape label: *pint.*

TUB: Use a **one-pound soft margarine tub**. These often hold 3 <u>standard cups</u>, as manufactured. Check, and cut to size if necessary.

Masking tape label: *tub.* (If tubs of other sizes are also circulating in your room, designate the official status of this particular tub by underlining the label.)

QUART: Cut a **liter bottle** to size so it holds 4 standard cups when filled fair and full.

Masking tape label: *quart.*

Label for Basic Materials

EXTRA CUPS
6 whole cups
1 half cup

EXTRA CUPS
6 whole cups
1 half cup

Store 6 whole cups (60 dram vials) and a half cup (30 dram vial) in a gallon storage jug. Apply label(s) with clear packaging tape. (SPECIAL NOTE: The 30 dram vial is slightly undersized. You may wish to <u>cut</u> a 60 or 40 dram vial <u>to size</u> as detailed on page 101.)

funny cups (measure 2)

puzzle books (measure 3, 5)

set of 5 measuring cups
(measure 3, 4, 5, 6, 7, 8)

E / MEASURE
special materials

CHAPTER SIGN: Glue to a 4 x 6 inch index card, and fold in half. Stand this sign in the space where you store special materials for this chapter. Or cut this sign in half, and glue both pieces to a grocery bag that has been cut to size, or to a box.

MEASURE
5 containers
quart
tub
pint
cup
half cup

MEASURE
5 containers
quart
tub
pint
cup
half cup

MEASURE
5 containers
quart
tub
pint
cup
half cup

Apply each label to a gallon storage jug with clear packaging tape. (Three duplicate sets recommended for large class sizes.)

Preparation: 1 copy

TEACHING NOTES

Purpose

To measure volume by counting single cups and by pouring into a calibrated bottle.

Introduction

▶ Fill all 4 cups fair and full, and stand them in a row. As you do this, remark that you are overfilling each cup, then giving it a single shake. (See teaching notes on pages 51 and 55.)

▶ Pour these 4 cups into the Measuring Bottle with scoop and funnel, without shaking down the contents. Hold up the bottle for your class to see. (Tilt it to level the the lentils, if necessary.) Notice that they reach to the 4-cup line.

▶ Now shake the bottle vigorously and tickle its sides. Notice that the lentils no longer reach the 4-cup line because they are more compact.

▶ How can we "undo" this compaction? Simply turn the bottle upside down, then rightside up. Tilt to level the lentils, and the bottle again reads 4 cups. Because this scale is calibrated for a loose pack, alway follow this invert-and-tilt-level procedure whenever you think the contents might have settled.

Focus

◆ How full will you fill the cups? *Fair and full.*

◆ When will you invert the bottle and tilt it level? *Whenever I think the lentils inside may have settled.*

◆ How will you read the Measuring Bottle? *See how high the lentils reach against the calibrated scale.*

◆ Can you predict in both directions:
  • How high a given number of cups will reach?
  • How many cups a given level will fill?

Checkpoint

◆ Fill _____ single cups fair and full. How high will they reach in the Measuring Bottle? Show me.

◆ Fill the Measuring Bottle with any amount of lentils. Invert and tilt level to make a loose pack. How many single cups will this fill? (Approximate predictions are OK. Small height variations in the wide Measuring Bottle produce larger height variations in the narrow cups.)

◆ May I see your Job Sheet?

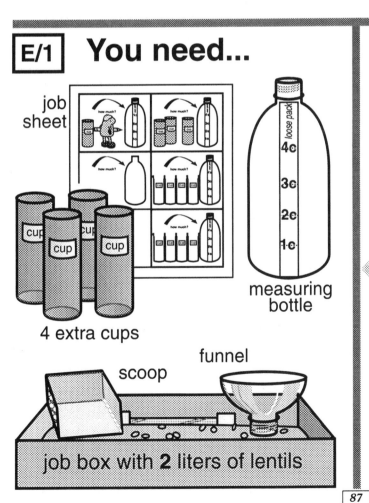

E/1 **You need...**

job sheet

4 extra cups

scoop

funnel

job box with **2 liters of lentils**

measuring bottle

measure 1

Pour cups.

cup cup cup cup

Draw the lentils.

how much?

cup

Draw how high.

how much?

cup

loose pack

4c
3c
2c
1c

Job Sheet: 1 copy per student

TEACHING NOTES

Purpose

To experience cups that have different heights and widths, but equal volume.

Introduction

▶ Hold up the 4 nested Funny Cups. Lift each one from its nest and read its name: tall cup, medium cup, short cup, very short cup.

▶ Order the cups by height; then by width. Notice that the tallest cup is also the narrowest; the shortest cup is also the widest.

▶ Point out that these special cups always belong together. Store them together whenever you put them away.

Focus

◆ Does each Funny Cup fill the Measuring Bottle to the same height?

◆ Can you pour a fair-and-full cup of lentils from one Funny Cup to the next, to the next, all around the circle?

◆ Can you trace around the mouth of each Funny Cup and label the circle?

Checkpoint

◆ Show me that these two Funny Cups hold the same amount. (Students might demonstrate this with the Measuring Bottle, or fill 1 cup fair and full and pour it into the empty.)

◆ Why does the short cup hold the same volume as the tall cup? *Because the short cup is wider than the tall cup. One cup of lentils doesn't pile as high in the short cup, because these lentils have more space to spread outward.*

◆ How high will lentils reach in the Measuring Bottle if you fill all the Funny Cups fair and full, then pour them all in? *To the 4-cup line.*

◆ Let me see your circle tracings. Why is the tall cup circle also the narrowest?

More

You drew a labeled *top* view of each cup. Now draw labeled *side* views.

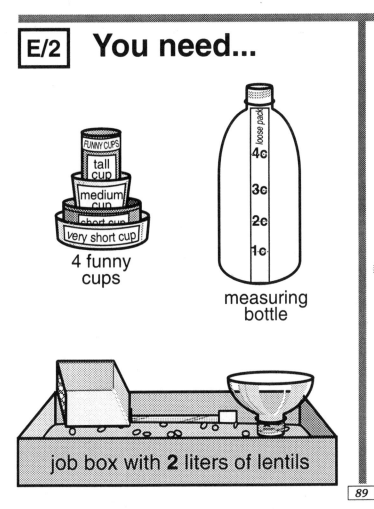

E/2 You need...

4 funny cups

measuring bottle

job box with **2** liters of lentils

measure 2

Pour funny cups.

Trace the top of each cup.

Label your drawings.

TEACHING NOTES

Purpose

To explore the quantitative relationship between cups, pints, "tubs" and quarts.

Introduction

Ask your class to watch carefully while you "count" the number of cups in a pint. Then proceed to add cups that are only partially full, counting up to some ridiculous number before the pint is finally fair and full. Announce that the pint holds, say, 9 cups!

If no one objects that the cups were only partially filled, repeat your "mistake," counting to an even higher, more absurd number. Lead your class to the realization that each cup must be filled fair and full to determine how many cups really do equal a pint. Ask a volunteer to demonstrate how many cups really fill a pint. Only 2!

Focus

◆ How many of these make one of those?

◆ How full will you make each cup you add?

◆ Can you solve the problems in this Puzzle Book?

Checkpoint

◆ How many _____ fill a _____? Prove it.

◆ Let me see you solve this equation in the Puzzle Book.

More

Will you get the same result if you use Funny Cups to fill the pint, tub or quart? *Yes. The cup's shape does not matter. Only its volume matters, the amount of lentils it holds.*

E/3 You need...

puzzle book

measure 3 puzzle book

cup cup cup cup = ?

MEASURE 5 containers

job box with **2 liters of lentils**

---

measure 3

Count!

whole cups

fair and full

cup

quart

Do the puzzle book.

measure 3 puzzle book

cup cup cup cup = ?

1

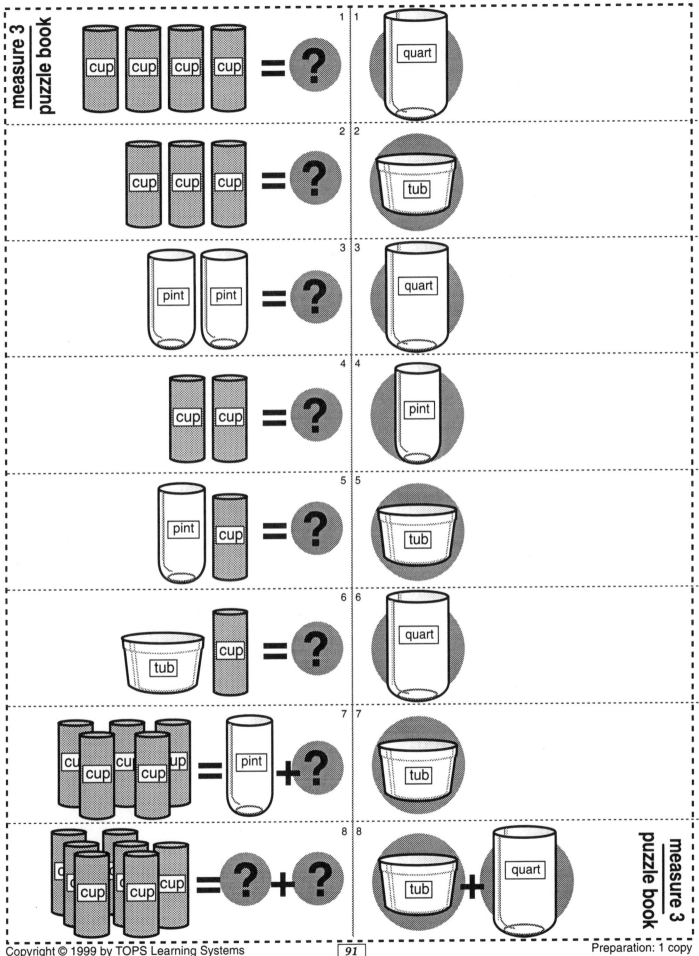

Preparation: 1 copy

measure 3
puzzle book

TEACHING NOTES

Purpose

To represent the relationship between cups, pints, "tubs" and quarts as written equalities.

Introduction

♦ Write a large equal sign on your blackboard. Ask a volunteer (student L) to stand to the left of this symbol and hold up 2 cups. Ask another (student R) to stand to the right holding 1 pint. Write the resulting equality on your blackboard: 2 cups hold the same as 1 pint.

♦ Now prove this relationship by pouring lentils. Take the empty cups from student L, fill them fair and full, and hand them back. Then ask student L to pour each cup into the pint held by student R. How full does the pint get? *Fair and full.*

Focus

◆ Where is the large equal sign? What does it mean?

◆ Can you use this equal sign to arrange an equality? What does it say? Can you prove it is true?

Checkpoint

◆ Let me see you arrange an equality and prove that it is true.

◆ May I see the equations you have written down? Can you read this one to me?

Special Note

Four cups (rather than just 1) enable beginners to arrange and copy 5 concrete sentences or equations:

2 cups hold the same as 1 pint.
3 cups hold the same as 1 tub.
4 cups hold the same as 1 quart.
1 cup and 1 pint hold the same as 1 tub.
1 cup and 1 tub hold the same as 1 quart.

Additional sentences are possible by adding number signs, but these have a more abstract character, since one container with multiplier represents many:

2 pint(s) hold the same as 1 quart.
5 cup(s) hold the same 1 pint and 1 tub.

DIRECTIONS:

1. Cut this page into 2 rectangles along the dashed lines.
2. Fold each rectangle in half, with numbers on the front and grey on the back.
3. Fold each in half again lengthwise to hide the grey inside.
4. Tape cut edges together along the entire length of each folded strip to make 2 very long stand-up signs.
5. Cut apart along the solid lines to make 13 stand-up signs.
6. Glue front and back layers together on each sign.

EQUATION TAG:

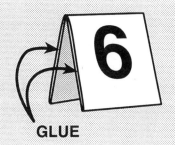

GLUE

hold the **same** as

= =

2 2

3 3

4 4

6 6

8 8

2 2

3 3

4 4

5 5

7 7

+ +

+ +

GLUE
6. Glue front and back layers together on each sign.
5. Cut apart along the solid lines to make 13 stand-up signs.
4. Tape cut edges together along the entire length of each folded strip to make 2 very long stand-up signs.
3. Fold each in half again lengthwise to hide the grey inside.
2. Fold each rectangle in half, with numbers on the front and grey on the back.
1. Cut this page into 2 rectangles along the dashed lines.

DIRECTIONS:

EQUATION TAG:

Preparation: 1 copy

TEACHING NOTES

Purpose

To understand half cups, cups, pints, "tubs" and quarts as simple multiples of each other.

Introduction

Line up a half cup, cup, pint and quart next to each other. Elicite a class response to each question, then pour lentils to verify.

Hold up the half cup. What's TWICE as big?
Hold up the cup. What's TWICE as big?
Hold up the pint. What's TWICE as big?
Hold up the quart. What's HALF as big?
Hold up the pint. What's HALF as big?
Hold up the cup. What's HALF as big?

Focus

◆ How many of these make one of those?

◆ Can you solve each problem in the Puzzle Book?

Checkpoint

◆ How many _____ fill a _____? Prove it.

◆ Hold up PAIRS of containers...
… related by 2 (also $1/2$).
half cup/cup; cup/pint; pint/quart
… related by 4 (also $1/4$).
half cup/pint; cup/quart
… related by 8 (also $1/8$).
half cup/quart

◆ Hold up PAIRS of containers...
… related by 3 (also $1/3$).
cup/tub
… related by 6 (also $1/6$).
half cup/tub

Special Note

Fair and full always refers to *loosely*-packed containers. Eight half-cups, for example, fill a quart fair and full, but will come up short if the quart is shaken down or "tickled." Gently correct these behaviors as you see them.

E/5 You need...

puzzle book

MEASURE
5 containers

job box with **2** liters of lentils

measure 5

Count!

half cups...

…fair and full

Do the puzzle book.

measure 5 puzzle book

how many?   half cup = cup

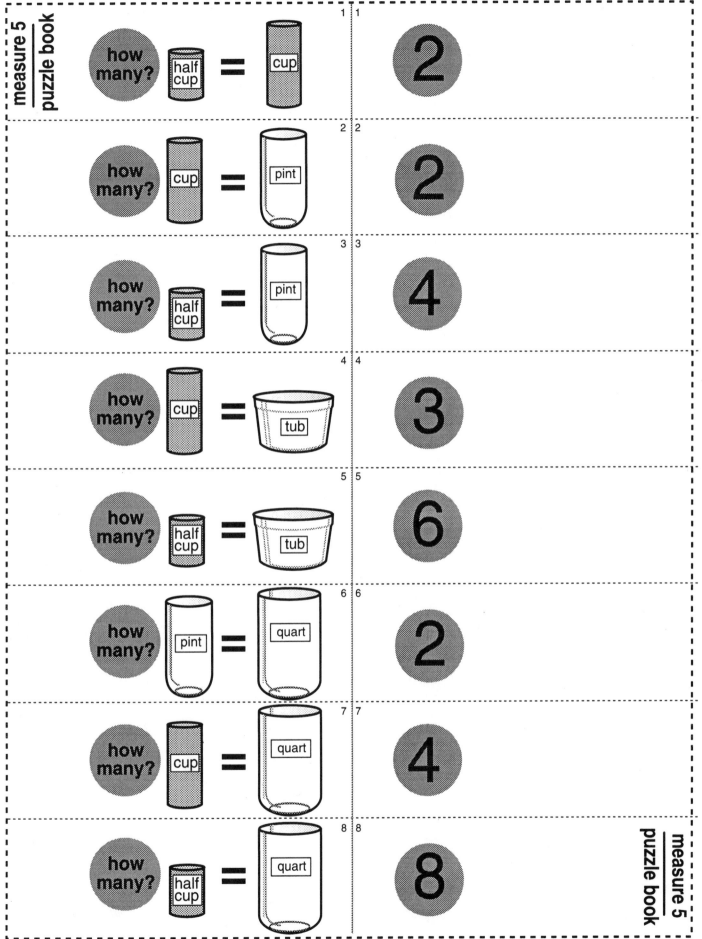

how many? half cup = cup 1 | 1 **2**

how many? cup = pint 2 | 2 **2**

how many? half cup = pint 3 | 3 **4**

how many? cup = tub 4 | 4 **3**

how many? half cup = tub 5 | 5 **6**

how many? pint = quart 6 | 6 **2**

how many? cup = quart 7 | 7 **4**

how many? half cup = quart 8 | 8 **8**

Preparation: 1 copy

TEACHING NOTES

Purpose

To explore and record the multiple relationships between half cups, cups, pints, "tubs" and quarts.

Introduction

♦ Set up an equality with Equation Tags and containers as illustrated in this Job Card. Write the number "1" on folded scratch paper. Display it after the equal sign to designate one cup. Point out that "1" is always understood, whether we use the tag or not. Recite the equation several times, with and without the "1" in place. Say it exactly the same way each time. Then crumple up the tag and throw it away. If "1" is always understood, tag or no tag, why bother using it at all?

♦ Write the illustrated equality in two ways. Point out that the "=" symbol is a short-hand way of saying "holds the same as." For example:

2 half cups hold the same as 1 cup
2 half cups = 1 cup

Focus

◆ Can you arrange / prove / write a true equality?

◆ Will you write "holds the same as," or use the equal symbol?

Checkpoint

◆ Model an equality and prove it is true.

◆ There is no number beside this container. What number is understood to be there, even though we can't see it? *The number "1."*

◆ May I see your written work?

E/6 You need...

equation tags

job box with **2 liters of lentils**

measure 6   👤 or 👥👥

Arrange.

Prove by pouring.

2 half cups hold the same as 1 whole cup.

Write.

TEACHING NOTES

Purpose

To express the volume of a half cup, cup, pint, "tub" and quart in terms of a calibrated measuring bottle.

Introduction

Sketch a large side view of the Measuring Bottle with cup calibrations on your blackboard. Announce that you will add lentils to this bottle (cumulatively) as follows:

First addition: 1 half cup
Second addition: 1 cup
Third addition: 1 pint
Fourth addition: 1 half cup

Ask volunteers to predict each new lentil level by drawing a light chalk line across your drawing. Then pour lentils to confirm the accuracy of each prediction.

Focus

◆ How many puzzles are on the Job Sheet? (8)

◆ Will you begin the first puzzle by pouring or drawing? (Pouring first is easier, since students see by experiment how high the lentils will rise. Drawing first exercises higher order thinking skills, requiring students to predict first before pouring.)

Checkpoint

◆ If ___ are filled fair and full, how high will the lentils reach when you pour them into this Measuring Bottle? (Point to any container or combination of containers.)

◆ May I see your Job Sheet?

More

Fill the Measuring Bottle with any unmeasured amount of lentils. Predict which container or combination of containers will be nearly full, fair and full, or overfull. (Remind students to invert the bottle of lentils and tilt level to maintain a loose pack.)

E/7 You need...

measuring bottle

MEASURE 5 containers

job sheet

job box with **2** liters of lentils

measure 7

Draw the lentils.

Check your answer.

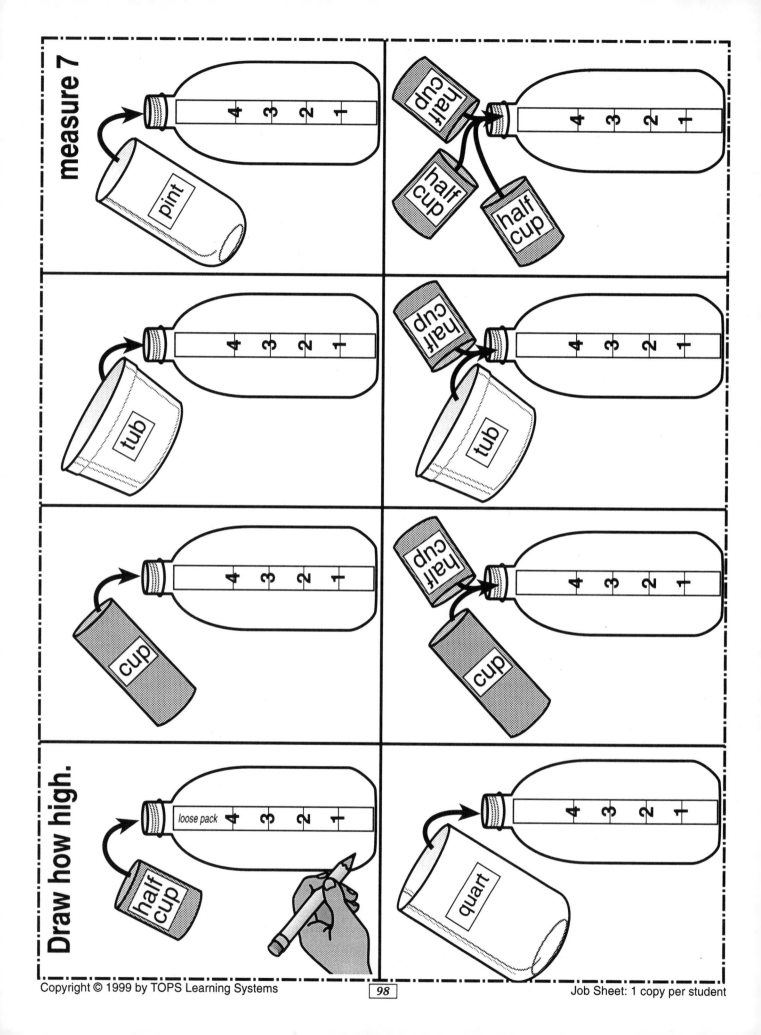

TEACHING NOTES

Purpose

To provide students with an approved way to pursue their own ideas about measuring.

Introduction

Display this card when you want to do the suggested activity or experiments that you design yourself.

Focus

◆ What do you want to study about measuring?

◆ Are these the materials you need?

Checkpoint

Report in words and pictures:
- What you did.
- What you learned.
- Questions you may still have.

Special Note

Stimulate new ideas: List the best ones on the opposite page!

E/8 You may use...

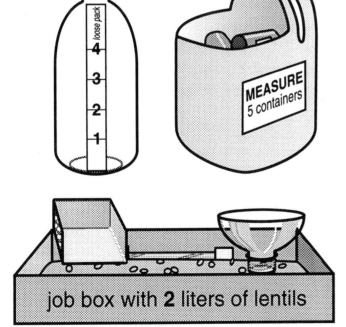

job box with **2 liters of lentils**

 Ask your teacher for other items. Tell why you need them.

measure 8 ☂ or ☂☂

On your own.

My idea:
How many hands full make a cup?

F / DIVIDE

In this chapter: Experience equality by setting equation tags between containers and pouring lentils across the equal sign. Solve the Puzzle Books. Fill a Prediction Tube with two of these or one of those. Do the parts you just poured in fill the whole? Lift the tube to verify. Divide geometric shapes. Write sums of fractions and add them up. Confirm your calculations by pouring.

job card **1**	count fair and full	104
job card **2**	solve the puzzle books	105
preparation	how many?	106
preparation	what's missing?	107
job card **3**	arrange equalities	108
job card **4**	prediction tube	109
job card **5**	divide and shade	110
preparation	fraction ruler	111
job sheet	divide each bar	112
job sheet	divide each shape	113
job card **6**	label, pour, measure, draw	114
preparation	fraction labels	115
job sheet	draw how high	116
job card **7**	how much all together?	117
job card **8**	on your own (keep dividing!)	118

Basic Materials: Quantities define maximums needed to support any one Job Card in this chapter. Store *high-quantity basics* (Job Boxes, liters of lentils, bottle lids, scoops, funnels) on and under a table or counter. Store *low-quantity basics* near the "basics" sign (see page 49) or in a "basics" box. Consult our Glossary on pages 6-10 for a full description of these items. See the next page for additional special materials used in this chapter.

- [ ] **1 job box**
- [ ] **up to 2 liters of lentils**
- [ ] **1 scoop**
- [ ] **1 funnel**
- [ ] **equation tags**
- [ ] **4 extra cups**
- [ ] **1 measuring bottle.**

Store these chapter-specific items together in a designated place. They require about 1/2 square foot of dedicated space. General classroom materials (like scissors and tape) are also listed below when used, while others (like pencil and paper) are always assumed.

set of 7 fraction cups (divide 1, 2, 3, 4, 6, 7, 8)

Store these containers in a **gallon storage jug**. You will need 3 identical sets of containers to serve a class of 30 students working simultaneously in all chapters of this curriculum. Photocopy the next two pages 3 times to obtain sufficient templates and labels.

2 WHOLE Cups: Use two **60 dram** <u>standard cups</u>.

Apply one label to each container.

ONE HALF: Use a **30 dram** <u>plastic vial</u>. Cut around the dotted lines of the small rectangular label and apply. (*SPECIAL NOTE:* This 30 dram half cup is 2 mm shorter than it should be. We used it without modification, and our primary students, who are not yet rocket scientists, never noticed the discrepancy. You might notice, however, and the fix is easy. Cut out the larger template that surrounds the half-cup label. Wrap it around a spare 40 or 60 dram vial, flush with the bottom, and tape it in place. This template (which is not quite square because the vial is not a perfect cylinder) now rises to the correct height. <u>Cut to size</u> with <u>toenail scissors</u>.

Remove the template. Cut out and apply the smaller label.)

ONE THIRD: Use a **20 dram plastic vial.**

Trim with toenail scissors to the *bottom edge* of the lock-tight nubs, so only very tiny bumps are left on the rim after the cut. Alternatively, use the template provided, following the directions above.

Apply the label.

ONE FOURTH: Use a **16 dram plastic vial.** Trim with toenail scissors through the *middle* of the lock-tight nubs. Alternatively, use the template provided, following the directions above.

Apply the label.

ONE FIFTH: Use a **16 dram plastic vial.** Use the corresponding template and cut to size.

Apply the label.

ONE SIXTH: Use a **16 dram or 13 dram plastic vial.** Use the corresponding template and cut to size.

Apply the label.

2 puzzle books (divide 2)

Photocopy these <u>booklets</u> and assemble as directed.

prediction tube (divide 4)

Use any long **cardboard tube** that fits inside a standard cup. Paper towel tube cores, measuring about 11 inches long, are suitable and don't need to be cut down. Or substitute two toilet tissue cores taped end to end. Cores from aluminum foil are generally made from more durable cardboard. A piece of 1¼ inch PVC pipe about 11 inches long is virtually indestructible.

Label with a marking pen: *prediction tube*.

fraction ruler (divide 5, 6, 8)

Copy the fraction ruler on page 111. Cut, fold and tape as directed.

fraction labels (divide 6, 7, 8)

1. Invert a **medium sized can** on your table. Place a **thick rubber band** around the outside, about three-fourths of the way up.

2. Slip 5 **craft sticks** between the rubber band and the can. Space them evenly around the circumference. Adjust their height so they stop about a finger-width above your table, and extend about the same distance above the rim of the inverted can.

3. Tape the lower ends of the sticks to the side of the can, using a long strip of masking tape all the way around.

4. Label the tops of the craft sticks, above the rubber band: halves, thirds, fourths, fifths and sixths.

5. Copy the Fraction Labels on page 115, and trim around the outside dotted line. Cut where indicated at each arrow to separate into 6 long strips.

6. Fold each strip in half along the grey line, and tape closed with a long strip of 2 inch, clear **packaging tape**, laminating them in the process.

7. Cut each strip into 5 labels. Sort by denomination into 5 piles of 6 each. Slide each pile behind the corresponding stick on the can.

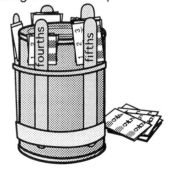

set of 7 fraction cups (divide 1, 2, 3, 4, 6, 7, 8)

2 puzzle books (divide 2)

prediction tube (divide 4)

fraction ruler (divide 5, 6, 8)

fraction labels (divide 6, 7, 8)

DIVIDE
7 containers

whole, whole
half
third
fourth
fifth
sixth

Apply to a gallon storage jug with clear packaging tape. Three duplicate sets of fraction cups are recommended for large class sizes. Please duplicate this page (and the next) 3 times for extra labels.

F / DIVIDE
special materials

CHAPTER SIGN: Glue to a 4 x 6 inch index card, and fold in half. Stand this sign in the space where you store special materials for this chapter. Or cut this sign in half, and glue both pieces to a grocery bag that has been cut to size, or to a box.

Apply this template to a **13 or 16 dram** vial.
Cut to the top of the paper.

$$\frac{1}{6}$$

sixth

Carefully cut out this "windshield" shape. Wrap it around the indicated vial, flush with the bottom, and tape in place. Cut the vial to the top of the template with toenail scissors. Remove the template, cut out the label on the dashed line, and tape it on the vial.

Preparation: 1 copy

Apply this template to a **40 or 60 dram** vial. Cut to the top of the paper.

$$\frac{1}{2}$$
half

$$\frac{1}{1}$$
whole

$$\frac{1}{1}$$
whole

Apply this template to a **20 dram vial**. Cut to the top of the paper.

$$\frac{1}{3}$$
third

Apply this template to a **16 dram** vial. Cut to the top of the paper.

$$\frac{1}{5}$$
fifth

Carefully cut out these "windshield" shapes. Wrap them around the indicated vial, flush with the bottom, and tape in place. Cut the vial to the top of the template with toenail scissors. Remove the template, cut out the label on the dashed line, and tape it on the vial.

Apply this template to a **16 dram** vial. Cut to the top of the paper.

$$\frac{1}{4}$$
fourth

Preparation: 1 copy

LESSON NOTES

Purpose

To explore volume relationships between whole cups and parts.

Introduction

◗ There are 2 whole cups and 5 fractional cups in the gallon jug. Hold up each size and read the label: one whole, one half, one third, one fourth, one fifth, one sixth. We always say the top number first, and bottom number last. Each bottom number is said in a special way: whole (not one), half (not two), third (not three), fourth (not four), fifth (not five), sixth (not six).

◗ Review fair and full. (See Lesson Notes on pages 51 and 55.)

◗ Count parts as you fill the whole. Say, for example: 1 third (fair and full), 2 thirds (fair and full), 3 thirds (fair and full), equal one whole. Keep repeating "fair and full" as long as necessary until the association is firmly established.

◗ Ask student volunteers to model the counting behavior you have just demonstrated.

◗ If your class thoroughly understands that 2 halves make a whole, count out some larger number of "halves" that are only partially filled, something absurd like, "These 5 halves fill the whole." Let your class correct this "mistake" by pointing out you "forgot" to fill them fair and full.

Focus

◆ How will you fill the parts? *Fair and full.*

◆ How can you keep from spilling? *Pour carefully or use the funnel.*

◆ What should you do if you spill? *Start over.*

Checkpoint

◆ Let me see you fill the _____ fair and full. (No patting down. No leveling off with fingers.)

◆ Let me hear you count how many of these parts fill this whole:
  • Are fair and full amounts used each time?
  • Can the student repeat each fraction correctly, saying the top part first and then the bottom? *One fourth, two fourths, three fourths, four fourths equal one whole.*

More

◆ Sort the containers from tallest to shortest; from largest to smallest.

◆ Write sentences: 1/2 overfills 1/3. 1/3 underfills 1/2.

F/1	**You need...**

scoop

funnel

job box with **2 liters of lentils**

divide 1   👤 or 👤👤

Scoop.   **Shake once.**

Count!

LESSON NOTES

Purpose

To solve equations by a hands-on process of trial and error.

Introduction

▶ Introduce Puzzle Book B: Copy the front-page equation across the bottom of your blackboard like this:

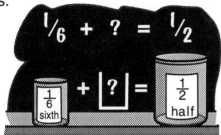

Select 1/4 as a possible solution by placing it on the chalk ledge in front of the question mark. Pour lentils. Show that 1/6 + 1/4 *underfills* 1/2.

Select 1/3 as a possible solution by placing it in front of the question mark. Show that 1/6 + 1/3 fills 1/2 fair and full. Turn the page in the Puzzle Book to find the answer on the back of page 1 and show that it "agrees" with this solution.

▶ Set up the next equation (on page 2) at the bottom of your blackboard:

$$1/4 + ? = 1/2$$

Use trial and error, as before, by first testing 1/5 (too little), then 1/3 (too much), then 1/4 (just right). Notice that the same 1/4 container needs to be used twice. It's OK to use the same container as many times as necessary.

Focus

◆ Which Puzzle Book will you try first? What will you do to solve the equations? *Find the named containers and start pouring lentils.*

◆ Can you guess any of the answers even before you pour the lentils?

Checkpoint

◆ Which Puzzle Book(s) did you try? Did you solve all the equations?

◆ Let me see you solve this equation.

More

Write your own equations and show that they are true.

F/2	**You need...**

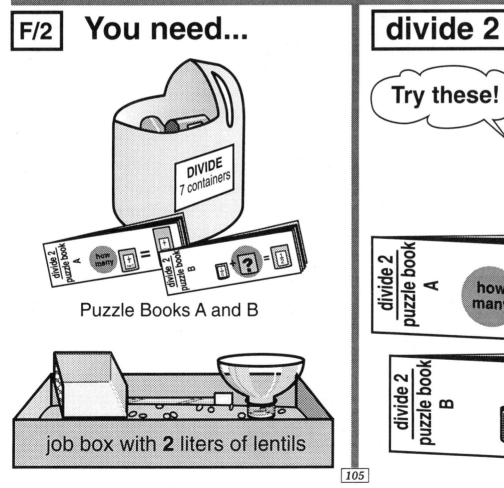

Puzzle Books A and B

job box with **2** liters of lentils

---

divide 2

Try these!

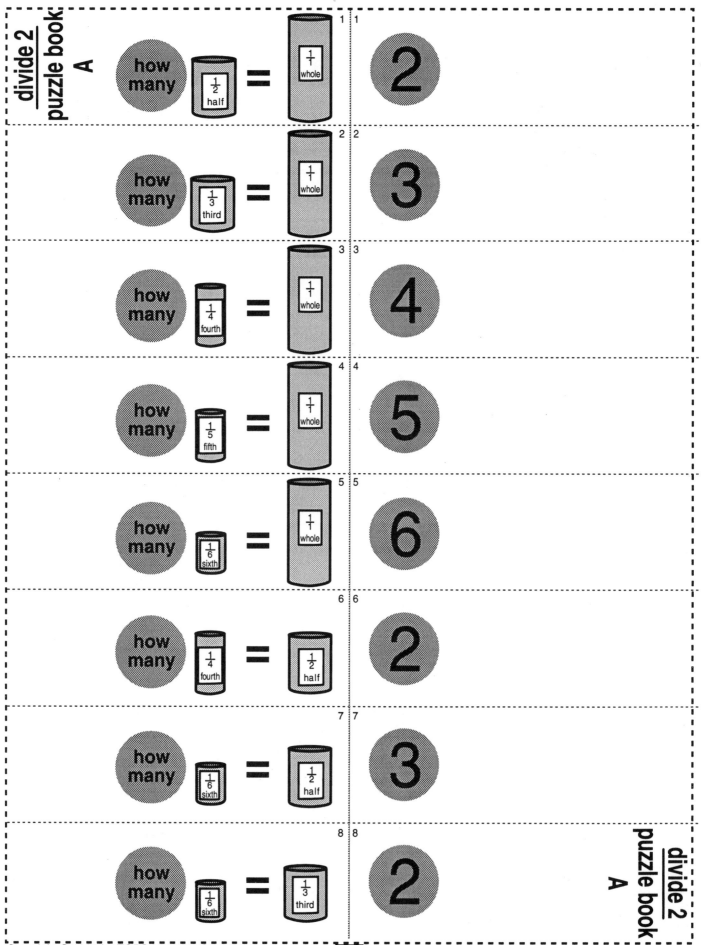

Preparation: 1 copy

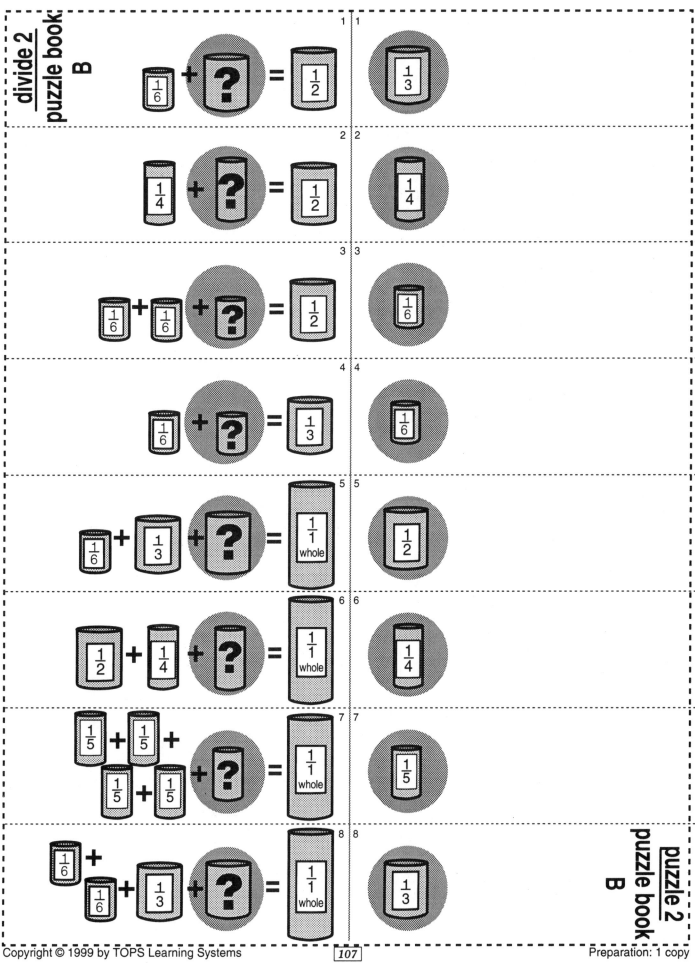

Preparation: 1 copy

LESSON NOTES

Purpose

To form concrete equations in the lentil box using containers and symbols. To write them down.

Introduction

◆ Set up an equation like this at the bottom of your blackboard. Ask what number should replace the question mark to make this a true equation. Pour lentils and count the parts to demonstrate that 5 is the answer: 1 fifth, 2 fifths, 3 fifths, 4 fifths, *5* fifths fills the whole.

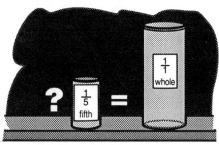

◆ Create another problem using the "+" sign for students to solve: ? + 1/3 = 1/2

Focus

◆ Set up a simple equation with the containers and Equation Tags. Then ask…
  • Can you pour lentils to prove this is true?
  • Can you write this equation down?

◆ Can you invent your own equations and write them down?

Checkpoint

◆ Let me see the equations you have written. Can you show me that this one is true?

◆ Can you arrange a true equation using a number tag and an equal tag? Using a number tag, a plus tag and an equal tag?

More

Can you form equations that equal some other container besides the whole?

2 fourths = 1 half
3 sixths = 1 half
2 sixths = 1 third
1 sixth + 1 third = 1 half

F/3 You need...

equation tags

DIVIDE 7 containers

job box with **2** liters of lentils

divide 3

Model an equation.

Pour it.

3 thirds is the same as 1 whole.

Use equation tags!

Write it down.

LESSON NOTES

Purpose

To review and consolidate a knowledge of basic fractions. To observe carefully.

Introduction

▶ Insert the Prediction Tube inside an empty whole cup. Tell your class to watch closely as you add any combination of parts. Ask by a show of hands how full the cup will be when you pull out the Prediction Tube: will it be underfull (thumb down), fair and full (thumb level) or overfull (thumb up)? Discuss reasons. Encourage debate. Then lift the tube to reveal the answer!

Use these combinations or others:
    underfull: 1/4 + 1/4 + 1/4
    fair and full: 1/2 + 1/4 + 1/4
    overfull: 1/2 + 1/3 + 1/3

▶ Repeat with the half cup:
    underfull: 1/5 + 1/5
    fair and full: 1/6 + 1/6 + 1/6
    overfull: 1/5 + 1/5 + 1/5

▶ So far, I have been pouring "fair." Is there a way of pouring that is unfair? *Yes, by NOT filling each part fair and full.*

Demonstrate an extreme example of this by underfilling the half with 20 or so very small portions, dumping each one into the prediction tube. Now is the cup fair and full? *Who knows? We can only guess. We can't use our brains.*

Focus

◆ Who will be the first pourer? The first observer?

◆ What should you do when the pourer doesn't add parts that are fair and full? *Start over.*

◆ How will you keep score? When will the game be over?

Checkpoint

◆ You can make 3 different predictions in this game. What are they? *overfull, fair and full, underfull*

◆ Watch carefully while I fill the Prediction Tube. How full is the cup?

◆ Let's play a round. Who pours first?

◆ Tell me about a turn that was tricky / interesting / surprising.

◆ How is predicting different from guessing? *A prediction requires careful looking and thinking. A guess requires no effort at all.*

F/4	**You need...**

prediction tube

Prediction Tube

DIVIDE
7 containers

job box with **2 liters of lentils**

---

divide 4

Predict right! Win a point!

I think I filled it fair and full.

I think you overfilled.

1/1 whole

LESSON NOTES

Purpose

To divide geometric figures into equal pieces, and shade some parts to represent fractions.

Introduction

▶ Ask students to recall from memory all 6 sizes of cups they have been working with. Write these across the top of your blackboard, in any order that students remember. Draw a large "pie" under each heading to represent each fraction.

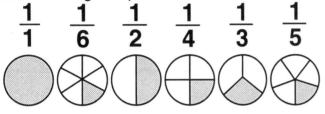

Notice that each fraction has a bottom number and a top number. The bottom number always says how many equal pieces we should divide the pie into. The top number tells how many of the pieces we get to eat. In these examples it seems we always eat just 1 of the pieces.

▶ Let's change the top number in some of the fractions so we can shade (eat) more pieces! (Erase numerators, substituting new numbers and shading the pies as students suggest. If

they request a numerator larger than a denominator, ask them to count the total number of pieces available, then go bake another pie.)

Can you think of 6 different ways to eat the whole pie? *1/1, 2/2, 3/3, 4/4, 5/5, 6/6*

Focus

◆ Look at Job Sheet 5A. Which section of the Fraction Ruler did the "peoplet" use to divide the first bar into 2 equal pieces? Why did she shade only one of the pieces?

◆ Look at the Job Sheet 5B. Why did the peoplet divide the square into 4 pieces, but only shade 3 of them?

Checkpoint

◆ What does the bottom of a fraction tell you? *The number of pieces to divide the whole into.*

◆ What does the top of a fraction tell you? *The number of pieces to shade.*

◆ Let me see your Job Sheets.

More

◆ Draw your own pies and divide them into any number of *equal* parts. "Eat" as many pieces as you like, then label what you ate with a fraction.

◆ Pour each fraction from Job Sheet 5A into a whole cup. Check that the lentils rise as high in the cup as you have drawn them on each bar.

F/5 You need...

2 job sheets

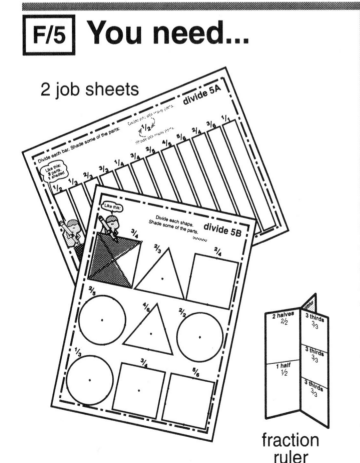

fraction ruler

divide 5

Divide and shade.

divide 5

Fraction Ruler

1. Cut out the rectangle on the narrow dashed line.
2. Fold like a fan on the solid lines.
2. Tape the left and right edges together (printing to the outside) to form a pleated tube.
3. Make a 3-page "book" of rulers: tape the pleats, back to back, at each end.

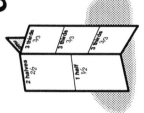

1 whole 1/1	2 halves 2/2	3 thirds 3/3	4 fourths 4/4	5 fifths 5/5	6 sixths 6/6
					5 sixths 5/6
			3 fourths 3/4	4 fifths 4/5	4 sixths 4/6
		2 thirds 2/3		3 fifths 3/5	3 sixths 3/6
			2 fourths 2/4	2 fifths 2/5	2 sixths 2/6
	1 half 1/2	1 third 1/3	1 fourth 1/4	1 fifth 1/5	1 sixth 1/6

Preparation: 1 copy

divide 5A

Divide each bar into equal parts. Shade some of the parts:

Divide into this many parts. **3/4** Shade this many parts.

1/1

3/6

2/4

5/6

4/5

2/5

3/4

1/4

3/3

2/3

1/3

Like this: **2** parts, **1** shaded

1/2

Job Sheet: 1 copy per student

Divide each shape.
Shade some of the parts.

divide 5B

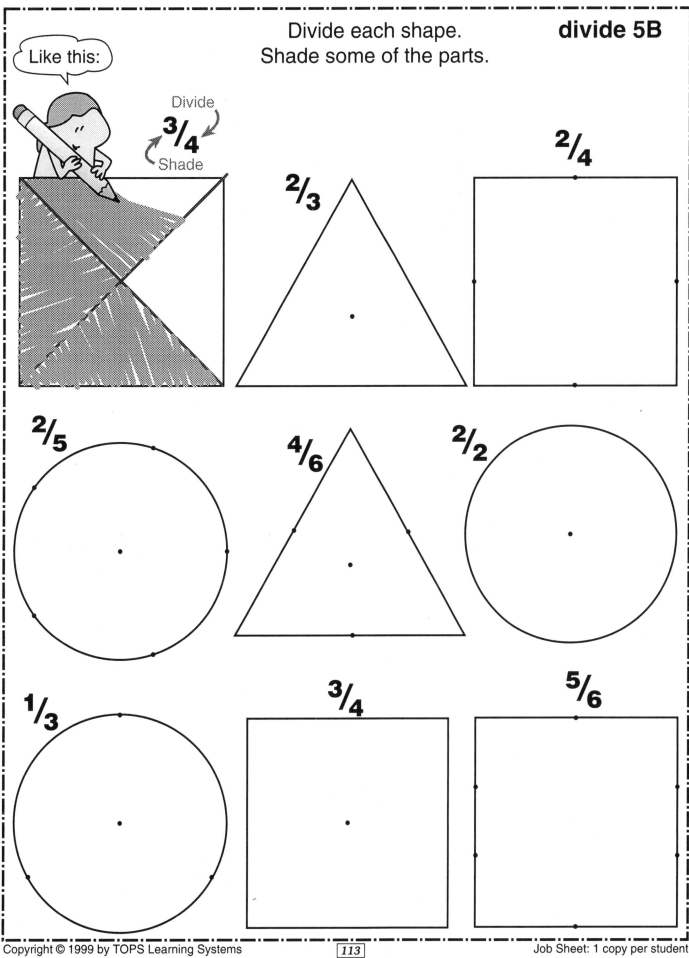

Like this:

Divide
3/4
Shade

2/3

2/4

2/5

4/6

2/2

1/3

3/4

5/6

Job Sheet: 1 copy per student

LESSON NOTES

Purpose

To choose fractions and express their meanings in terms of pouring, measuring and drawing.

Introduction

▶ Hold up the can that organizes the Fraction Labels. Notice that there are 5 groups of labels (halves, thirds, fourths, fifths, sixths) with 6 identical copies in each group.

▶ Select a label from one group. Notice that it has fractions on both sides. Choose any one fraction to insert behind the paper clip on a whole cup.

▶ The paper clip now points to one particular fraction. Who can read it and pour that many lentils into the whole cup?

▶ Measure the level of the lentils in the whole cup with a Fraction Ruler. Be sure to use the correct part of the ruler for the fraction you choose.

▶ Using that line on the ruler, shade the first cup on the Job Sheet to the correct height. Be sure to record this fraction in the box above.

▶ We have filled the first cup on the Job Sheet. What fraction shall we choose next? One that fills the next cup higher? Lower? By the same amount?

Focus

◆ Different fractions are written on this label. Can you slide the one you want behind the paper clip?

◆ May I suggest a pattern you might like to pour, measure and draw?

Stair Steps:
- 1/6, 2/6, 3/6, 4/6, 5/6, 6/6. • 1/6, 1/5, 1/4, 1/3, 1/2, 1/1.
- 5/5, 4/5, 3/5, 2/5, 1/5, 0/5. • 2/2, 2/3, 2/4, 2/5, 2/6.
- 1/3, 2/3, 3/3, 3/4, 2/4, 1/4.

Flat Floors:
- 6/6, 5/5, 4/4, 3/3, 2/2, 1/1 • 1/2, 2/4, 3/6. • 1/3, 2/6.

Note: 1/1 equals a full cup. 0/5 equals an empty cup. If students ask for these labels, suggest they write their own on scratch paper.

◆ How will you record this pattern? *I'll use a Job Sheet and Fraction Ruler. And I'll label each box.*

Checkpoint

◆ Let me see your Job Sheet drawings. Tell me what you did.

◆ I see you have made stair steps. How might you make the steps steeper / more shallow / go down / go up?

◆ I see you have drawn the fraction _____. Can you think of a different fraction that looks exactly the same?

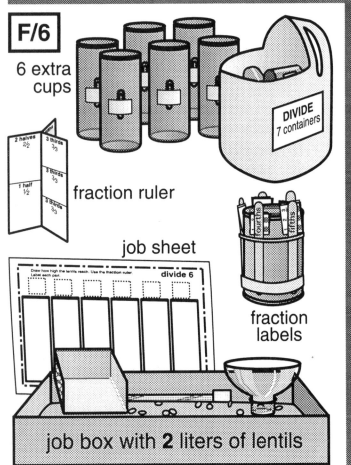

F/6

6 extra cups

DIVIDE 7 containers

fraction ruler

2 halves 2/2   3 thirds 3/3   3 thirds 3/3   1 half 1/2   3 thirds 3/3

job sheet

Draw how high the lentils reach. Use the fraction ruler. Label each part.   divide 6

fraction labels

job box with **2 liters of lentils**

divide 6   🧍 or 🧍🧍

Label and pour.

1 sixth, 2 sixths, 3 sixths...

1/6   1/6 2/6 3/6   1/6 2/6 3/6   1/6

Measure and draw.

6 sixths 6/6
5 sixths 5/6
4 sixths 4/6
3 sixths 3/6
2 sixths 2/6
1 sixth 1/6

Draw how high the lentils reach. Use the fraction ruler. Label each part.   divide 6

divide 6

| $\frac{1}{2}$ | $\frac{1}{3}$ | $\frac{2}{3}$ | $\frac{1}{4}$ | $\frac{2}{4}$ | $\frac{1}{5}$ | $\frac{2}{5}$ | $\frac{3}{5}$ | $\frac{1}{6}$ | $\frac{2}{6}$ | $\frac{3}{6}$ |

$\frac{2}{2}$ $\frac{3}{3}$ $\frac{4}{4}$ $\frac{3}{4}$ $\frac{5}{5}$ $\frac{4}{5}$ $\frac{6}{6}$ $\frac{5}{6}$ $\frac{4}{6}$

| $\frac{1}{2}$ | $\frac{1}{3}$ | $\frac{2}{3}$ | $\frac{1}{4}$ | $\frac{2}{4}$ | $\frac{1}{5}$ | $\frac{2}{5}$ | $\frac{3}{5}$ | $\frac{1}{6}$ | $\frac{2}{6}$ | $\frac{3}{6}$ |

$\frac{2}{2}$ $\frac{3}{3}$ $\frac{4}{4}$ $\frac{3}{4}$ $\frac{5}{5}$ $\frac{4}{5}$ $\frac{6}{6}$ $\frac{5}{6}$ $\frac{4}{6}$

| $\frac{1}{2}$ | $\frac{1}{3}$ | $\frac{2}{3}$ | $\frac{1}{4}$ | $\frac{2}{4}$ | $\frac{1}{5}$ | $\frac{2}{5}$ | $\frac{3}{5}$ | $\frac{1}{6}$ | $\frac{2}{6}$ | $\frac{3}{6}$ |

$\frac{2}{2}$ $\frac{3}{3}$ $\frac{4}{4}$ $\frac{3}{4}$ $\frac{5}{5}$ $\frac{4}{5}$ $\frac{6}{6}$ $\frac{5}{6}$ $\frac{4}{6}$

| $\frac{1}{2}$ | $\frac{1}{3}$ | $\frac{2}{3}$ | $\frac{1}{4}$ | $\frac{2}{4}$ | $\frac{1}{5}$ | $\frac{2}{5}$ | $\frac{3}{5}$ | $\frac{1}{6}$ | $\frac{2}{6}$ | $\frac{3}{6}$ |

$\frac{2}{2}$ $\frac{3}{3}$ $\frac{4}{4}$ $\frac{3}{4}$ $\frac{5}{5}$ $\frac{4}{5}$ $\frac{6}{6}$ $\frac{5}{6}$ $\frac{4}{6}$

| $\frac{1}{2}$ | $\frac{1}{3}$ | $\frac{2}{3}$ | $\frac{1}{4}$ | $\frac{2}{4}$ | $\frac{1}{5}$ | $\frac{2}{5}$ | $\frac{3}{5}$ | $\frac{1}{6}$ | $\frac{2}{6}$ | $\frac{3}{6}$ |

$\frac{2}{2}$ $\frac{3}{3}$ $\frac{4}{4}$ $\frac{3}{4}$ $\frac{5}{5}$ $\frac{4}{5}$ $\frac{6}{6}$ $\frac{5}{6}$ $\frac{4}{6}$

| $\frac{1}{2}$ | $\frac{1}{3}$ | $\frac{2}{3}$ | $\frac{1}{4}$ | $\frac{2}{4}$ | $\frac{1}{5}$ | $\frac{2}{5}$ | $\frac{3}{5}$ | $\frac{1}{6}$ | $\frac{2}{6}$ | $\frac{3}{6}$ |

$\frac{2}{2}$ $\frac{3}{3}$ $\frac{4}{4}$ $\frac{3}{4}$ $\frac{5}{5}$ $\frac{4}{5}$ $\frac{6}{6}$ $\frac{5}{6}$ $\frac{4}{6}$

CUT CUT CUT CUT CUT CUT

Preparation: 1 copy

divide 6

Draw how high the lentils reach. Use the **fraction ruler**. Label each part.

Job Sheet: 1 copy per student

LESSON NOTES
Purpose

To pour out fractional parts and experience how they add together. To predict and record simple fractional sums.

Introduction

▶ Use the half cup to fill 6 whole cups halfway. (This is 2 more than shown on the Job Card.) Label each with Fraction Labels as illustrated.

▶ Discuss how high all these lentils might reach when poured into the Measuring Bottle: Who predicts 1 cup? 2 cups? 3 cups? 4 cups? Why do you think so? Slide the rubber band on the bottle to the predicted level. Write predictions on scratch paper to slide underneath. Different predictions are fine, just add more rubber bands.

▶ Watch the lentils rise, half cup by half cup, as you funnel each container into the Measuring Bottle. Point out that you have avoided "tickling" or shaking the bottle, since it must keep a loose pack. Summarize the result on your blackboard:

$$1/2 + 1/2 + 1/2 + 1/2 + 1/2 + 1/2 = 3$$

▶ Settle the 3 cups of lentils below the 3c mark by shaking and "tickling." Show how to undo this settling by inverting the bottle and tilting to level the contents. Notice how the lentils once again reach the 3c mark.

Focus

◆ What can you do with the halves? How many cups will you label and fill halfway? (Encourage students to hone their predicting and recording skills with the halves before moving on to fourths, thirds, sixths and fifths, in that order.)

◆ What question should you ask yourself after you have filled and labeled the cups you want to use? *How much altogether?*

◆ May I suggest a problem you might like? Some recommended sums (simple to complex):

$$1/2 + 1/2 = 1$$
$$1/2 + 1/2 + 1/2 = 1\ 1/2$$
$$1/4 + 1/4 + 1/4 + 1/4 = 1$$
$$1/4 + 1/4 + 1/4 + 1/4 + 1/4 = 1\ 1/4$$
$$1/3 + 2/3 = 1$$
$$1/3 + 2/3 + 3/3 = 2$$
$$1/6 + 1/6 + 1/6 + 1/6 + 1/6 + 1/6 = 1$$
$$1/6 + 2/6 + 3/6 + 4/6 + 5/6 + 6/6 = 3\ 1/2$$
$$1/5 + 1/5 + 1/5 + 1/5 + 1/5 = 1$$
$$3/5 + 3/5 + 3/5 + 3/5 + 3/5 = 3$$
$$1/2 + 2/4 + 1/3 + 4/6 = 2$$

Checkpoint

◆ Let me see you pour a problem using the halves / fourths / thirds / sixths / fifths.

◆ May I see your written work?

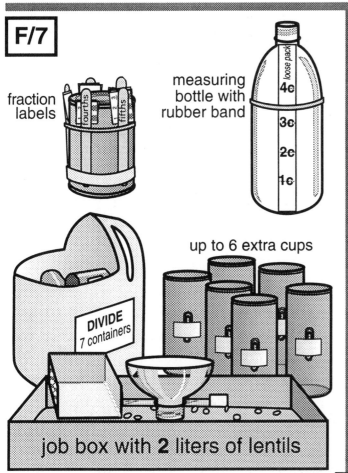

F/7

fraction labels

measuring bottle with rubber band

loose pack
4c
3c
2c
1c

up to 6 extra cups

DIVIDE
7 containers

job box with **2 liters of lentils**

divide 7　👤 or 👤👤

Label and partly fill.

Predict how much.

loose pack
4c
3c
2c
1c

I think it will be 2 cups.

Record.

TEACHING NOTES

Purpose

To provide students with an approved way to pursue their own ideas about dividing or combining quantities of lentils.

Introduction

Display this card whenever you want to do this suggested activity or experiments that you design yourself.

Focus

◆ What do you want to study about dividing lentils?

◆ Are these the materials you need?

Checkpoint

Report in words and pictures:
• What you did.
• What you learned.
• Questions you may still have.

Special Note

Recognize excellence: Record some of your students' most creative ideas on the opposite page.

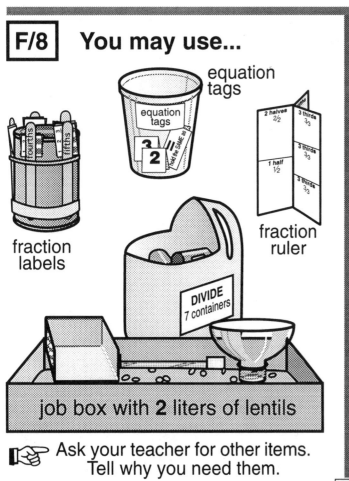

G / CALIBRATE

In this chapter: Calibrate a liter bottle with rubber bands, cup by cup, half cup by half cup. Decipher an "ancient Egyptian" code. Compare how high a cup of lentils rises in fat and skinny containers. Graph height against volume in both bar and grid formats. Notice how the size and shape of a container determines the steepness of the graph line; how graph lines become steeper or shallower as the shape of the bottle changes.

job card **1**	calibrate in cups	122
job card **2**	decode the hieroglyphics	123
job card **3**	calibrate in fractions	124
job card **4**	measure the height of one cup	125
job sheet	bar graph: changing height	126
job card **5**	calibrate a craft stick	127
job sheet	bar graph: increasing height	128
job card **6**	calibrate a paper strip	129
job sheet	line graph: increasing height	130
preparation	calibrating strips	131
job card **7**	calibrate a masking tape strip	132
job sheet	plot different graph lines	133
job card **8**	on your own (all-in-one bottle)	134

Basic Materials: Quantities define maximums needed to support any one Job Card in this chapter. Store *high-quantity basics* (Job Boxes, liters of lentils, bottle lids, scoops, funnels) on and under a table or counter. Store *low-quantity basics* near the "basics" sign (see page 49) or in a "basics" box. Consult our Glossary on pages 6-10 for a full description of these items. See the next page for additional special materials used in this chapter.

- [ ] **1 job box**
- [ ] **up to 2 liters of lentils**
- [ ] **1 scoop**
- [ ] **1 funnel**
- [ ] **1 extra cup and half cup**
- [ ] **up to 3 tubs**
- [ ] **up to 6 clear cups**
- [ ] **4 nesting containers**
- [ ] **2 craft sticks**
- [ ] **1 paper plate**
- [ ] **baby food jars (small, medium, large)**
- [ ] **1 half liter**

Store these chapter-specific items together in a designated place. They require about 1 square foot of dedicated space. General classroom materials (like scissors and tape) are also listed below when used, while others (like pencil and paper) are always assumed.

cup bands (calibrate 1)

1. Tape 2 **craft sticks** to the side of an inverted <u>medium-sized can</u> with **packaging tape** so they stick above the can about a third of their total length. Label them "wholes" and "halves" with a **permanent marker**.

2. Label ³/₄ inch masking tape while it is still on the roll:

½ cup	½ cup	1 cup	1 cup	1½ cup	1½ cup	2 cups	2 cups	2½ cups	2½ cups	3 cups	3 cups	3½ cups	3½ cups	4 cups	4 cups

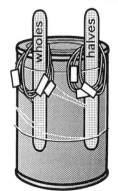

3. Cut off each double label, and fold it over a **thin rubber band**. Store the whole number bands over the "whole" craft stick; the mixed number bands over the "halves" craft stick.

cloth bundle (calibrate 2)

Copy and cut out the "Egyptian hieroglyphic" labels. Apply the 3 smaller labels to dedicated <u>baby food jars</u> with clear **packaging tape** as follows:

    BUG: small BFJ;

    BIRD: medium BFJ;

    GOAT: large BFJ.

Apply the tall label to a dedicated <u>standard cup</u>, keeping its bottom edge flush with the bottom of the cup. Wrap these 4 objects in a **towel or cloth** that looks dingy and old.

fraction bands (calibrate 3)

1. Evenly space 5 **craft sticks** around an inverted <u>medium-sized can</u> so they project above it about a third of their length. Hold them there with 2 **rubber bands**, spaced as far apart as possible.

2. Tape between these rubber bands with **packaging tape** all the way around. Then remove the rubber bands.

3. Label the top of each craft stick with these fractions around the can, using a **fine-tipped marker** or **pen**: *halves, thirds, fourths, fifths, sixths.*

4. Label ³/₄ inch masking tape, still on the roll, with a marking pen. Write all fractions in vertical format, with a period following each one:

$$\frac{1}{2}. \quad \frac{1}{2}. \quad \frac{2}{2}. \quad \frac{2}{2}. \longrightarrow$$

¹/₂. ¹/₂. ²/₂. ²/₂.

¹/₃. ¹/₃. ²/₃. ²/₃. ³/₃. ³/₃.

¹/₄. ¹/₄. ²/₄. ²/₄. ³/₄. ³/₄. ⁴/₄. ⁴/₄.

¹/₅. ¹/₅. ²/₅. ²/₅. ³/₅. ³/₅. ⁴/₅. ⁴/₅. ⁵/₅. ⁵/₅.

¹/₆. ¹/₆. ²/₆. ²/₆. ³/₆. ³/₆. ⁴/₆. ⁴/₆. ⁵/₆. ⁵/₆. ⁶/₆. ⁶/₆.

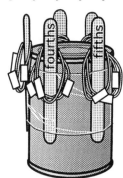

5. Cut off each double label (each duplicated fraction) and fold it over a thin rubber band. Hang these Fraction Bands over the corresponding craft sticks.

large and small milk cartons (calibrate 5, 6)

<u>Cut</u> a **quart and half-gallon milk carton** <u>to size</u>, equal in height to the length of a craft stick.

calibrating strips (calibrate 6)

Copy the line master on page 131. Separate with **scissors or a paper cutter** and store in a <u>clear cup</u>.

Attach the label supplied on the next page.

clear tape (calibrate 6)

masking tape (calibrate 7)

blunt scissors (calibrate 7, 8)

cup bands (calibrate 1)

cloth bundle (calibrate 2)

fraction bands (calibrate 3)

large and small milk cartons
(calibrate 5, 6)

calibrating strips (calibrate 6)

G / CALIBRATE
special materials

CHAPTER SIGN: Glue to a 4 x 6 inch index card, and fold in half.
Stand this sign in the space where you store special materials for
this chapter. Or cut this sign in half, and glue both pieces to a
grocery bag that has been cut to size, or to a box.

calibrate 6
calibrating
strips

Apply to <u>clear cup</u> with clear
packaging tape.

See instructions for Cloth Bundle,
previous page.

Preparation: 1 copy

TEACHING NOTES

Purpose

To calibrate and label a liter bottle in cups and half cups.

Introduction

◗ Model how to calibrate (mark) the liter bottle in cups: Pour in 1 cup filled fair and full; mark the lentil level with the 1-cup band; repeat with 2 cups, 3 cups and 4 cups.

◗ Shake the bottle and "tickle" its sides to show how 4 cups of lentils can pack together and look like less than 4 cups. Invert the bottle (to "fluff" the lentils) then tilt far enough to level them again at the 4-cup band.

◗ Empty the bottle, leaving the Cup Bands in place. Ask if anyone can measure 2 cups of lentils *in the bottle*. After a student fills it to the 2-cup line, confirm that it really does fill 2 whole cups, *approximately*. (Expect significant error. Even with well placed markers, small height variations in the wide bottle produce larger height variations in the narrower cups.)

◗ Empty the bottle again, leaving the Cup Bands in place. Refill the bottle with a cup filled *much less* than fair-and-full. Prompt students to explain why the lentils fail to reach the 1-cup band.

◗ Once students understand cup calibrations, allow them to discover half-cup calibrations on their own. The only preparation you might give them is an introduction to mixed numbers. Write these in a random series, then order them in a vertical column from least on the bottom to most on top: 1, 3, 2, 4, ½, 2½, 1½, 3½.

Focus

◆ Can you fill this cup fair and full, then mark how high the lentils reach?

◆ Show me that your calibrations really do measure cups / half cups.

◆ Can you calibrate cups and half cups using only the half cup? Using the half cup only once?

Checkpoint

◆ Are the rubber bands level all the way around and reasonably spaced?

◆ Show me how to measure _____ cups using this bottle you have calibrated.

◆ If 1½ cups are in the bottle, and you add 1 cup more, how high will the lentils reach?

More

Draw your calibrated bottle full size. Lay it on its side to outline the bottle and accurately position the rubber bands.

G/1 You need...

cup bands

extra cup and extra half cup

liter bottle

scoop

funnel

job box with **2 liters of lentils**

calibrate 1 ⮚ or ⮚⮚

Calibrate.

...pour...

...mark.

2 cups

1½ cups

1 cup

½ cup

2½ cups

TEACHING NOTES

Purpose

To solve a calibration mystery. To imagine and invent.

Introduction

Review this vocabulary, then read the story:
- Archeologist: A scientist who searches for clues about people who lived long ago.
- Pharaoh: An ancient Egyptian king.
- sarcophagus: A stone coffin.
- hieroglyphics: A language written in pictures.

Let's pretend that we are archeologists, exploring the cool, dark tomb of an Egyptian Pharaoh. Inside the sarcophagus we discover a very old Cloth Bundle enclosing 4 mystery objects. Let's look inside!

What's this? A tall vessel with strange hieroglyphics. And 3 pots of different sizes that look amazing similar to modern baby food jars! What kinds of marks do you see on these objects?

Your mission, should you choose to accept it, is to crack the code and discover what these pictures are trying to tell us.

Focus

◆ What do you suppose these markings mean? Can you crack the code?

◆ Try pouring a short pot into the tall cup. Hmmmm.

◆ Can you write a story about these cups? What might they say about the people who used them?

◆ Can you write directions, telling someone else how to use these cups?

◆ Can you draw an outline of these cups on paper and make your own symbols and decorations?

Checkpoint

◆ Why are there 2 bugs on the tall cup, but only 1 bird and 1 goat?

◆ What do these symbols mean? Pour lentils to explain your theory.

◆ May I see your picture/written work? Please tell me about it.

More

Invent your own picture writing. Create a secret message.

G/2	You need...

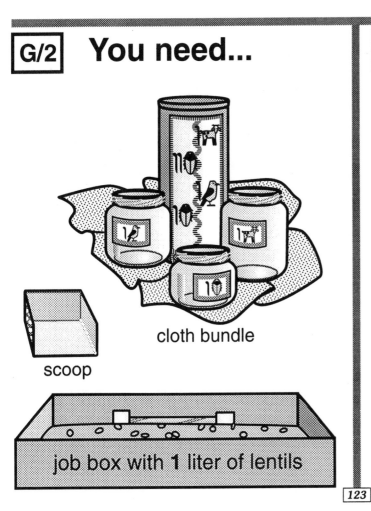

cloth bundle

scoop

job box with **1** liter of lentils

---

calibrate 2 🧍 or 🧍🧍

What can all this mean?

TEACHING NOTES

Purpose

To calibrate a liter bottle by first pouring out all the lentils in equal portions, then pouring each portion back, marking the level with rubber bands.

Introduction

‣ You want to divide a whole liter of lentils into 3 equal portions.

• How full will you fill the liter bottle? *Fair and full; to the top with a loose pack.*

• Will you use these tubs or clear cups? *Use the tubs. Three clear cups aren't big enough to hold a liter of lentils.*

‣ Divide the lentils as if you were dividing dessert.

• What part of the whole bottle does each tub contain? *1/3 part.*

• How many parts altogether? *1/3, 2/3, 3/3 equal 1 whole.*

‣ Show your class the Fraction Bands. Notice how they are organized into 2 halves, 3 thirds, 4 fourths, 5 fifths and 6 sixths: Which group of rubber bands should we use to calibrate the bottle with the portions we have just poured? *The thirds.*

‣ Use a funnel to pour each third back, one at a time. After each addition, calibrate how high the lentils reach with a Fraction Band. Deliberately put them on in the wrong order and read the calibrations: 2/3, 3/3, 1/3. Is this right? *Nooooo.*

‣ Place the Fraction Bands in the correct 1/3, 2/3, 3/3 order. Do this by temporarily marking each position with a small piece of masking tape.

Focus

◆ Into how many equal parts will you divide the whole liter: 2, 3, 4, 5, or 6? Will you use tubs or clear cups to hold all the lentils?

◆ Which group of rubber bands will you use to calibrate the bottle? Which particular rubber band will you apply first?

◆ What if you spill some of the lentils? *Refill the bottle and start over.*

Checkpoint

◆ Show me your calibrated bottle.

• Let me hear you count the parts.

• Use it to pour _____ liter of lentils into this tub. (Choose fractions thoughtfully, that challenge but don't overwhelm.)

◆ Show me your drawing. Did you label everything?

◆ Can you draw the liter again with different calibrations?

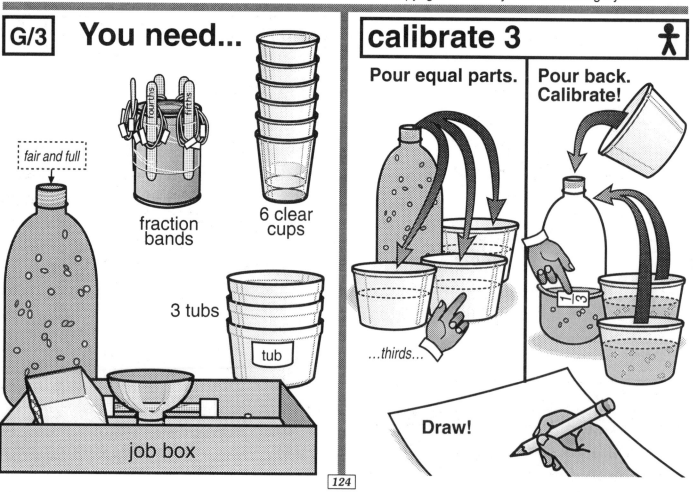

G/3 **You need...** fraction bands — 6 clear cups — 3 tubs — tub — fair and full — job box

calibrate 3 — Pour equal parts. ...thirds... — Pour back. Calibrate! — 1/3 — Draw!

TEACHING NOTES

Purpose

To experience how one cup fills containers of different widths to different heights. To make a bar graph of this relationship.

Introduction

♦ Who can sort these containers by width? Let's start with the narrowest and end with the widest.

♦ I'd like to work with a container that is not too narrow and not too wide. Which one should I choose? *The medium container.*

♦ Who would like to pour a cup of lentils into this medium container for me. Is the cup fair and full?

♦ How high does 1 cup of lentils reach in this medium container? I'll find out by pushing a craft stick all the way to the bottom, then slide my thumb down the stick to the top of the lentils. The lentils reach this high!

♦ Watch me graph this height on this bar graph. (Ask a helper to hold up the Job Sheet, or tape it to a wall.)

• Where should I draw my bar? (Read each of the labels at the bottom of the graph out loud.)

• Watch me rest this stick on the bottom line, then mark where my thumb is. Can you make your own graph like this?

Focus

◆ Which container will you graph first? What amount will you pour into it? *One cup, fair and full.*

◆ How will you measure the height of the lentils?

◆ Is it important to push the craft stick all the way to the bottom of the carton? *Yes.* (Otherwise, the calibration mark will be too low on the stick.)

◆ Where will you draw the bar for this container on your graph? (The last item on this graph, the cup, will exceed the length of the craft stick. Students should use the actual cup, like a ruler, to measure how high to draw the bar.)

Checkpoint

◆ Show me your bar graph. Why is this bar longer than than one?

◆ Why does your bar graph start out low, but end up high? *Because the containers start out wide, but end up narrow.*

◆ Why do lentils rise to a low level in a wide container? *Because they spread out, not up.*

◆ Why do lentils rise to a high level in a narrow container? *Because they pile up, not out.*

◆ How tall is 1 cup of lentils? *It depends on the width of the container you pour it into.*

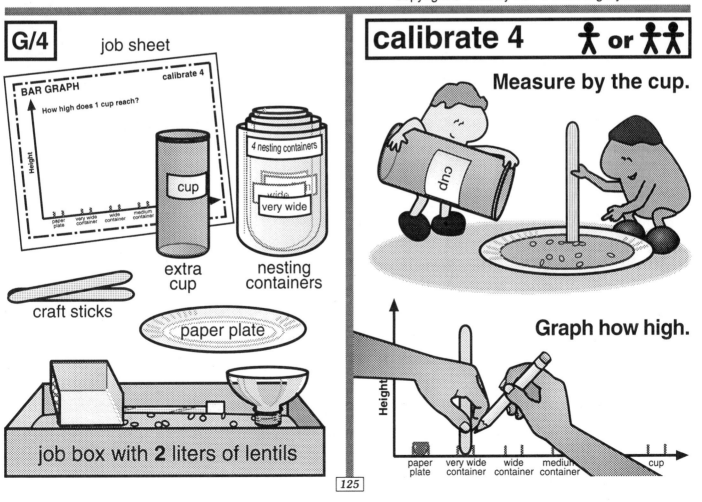

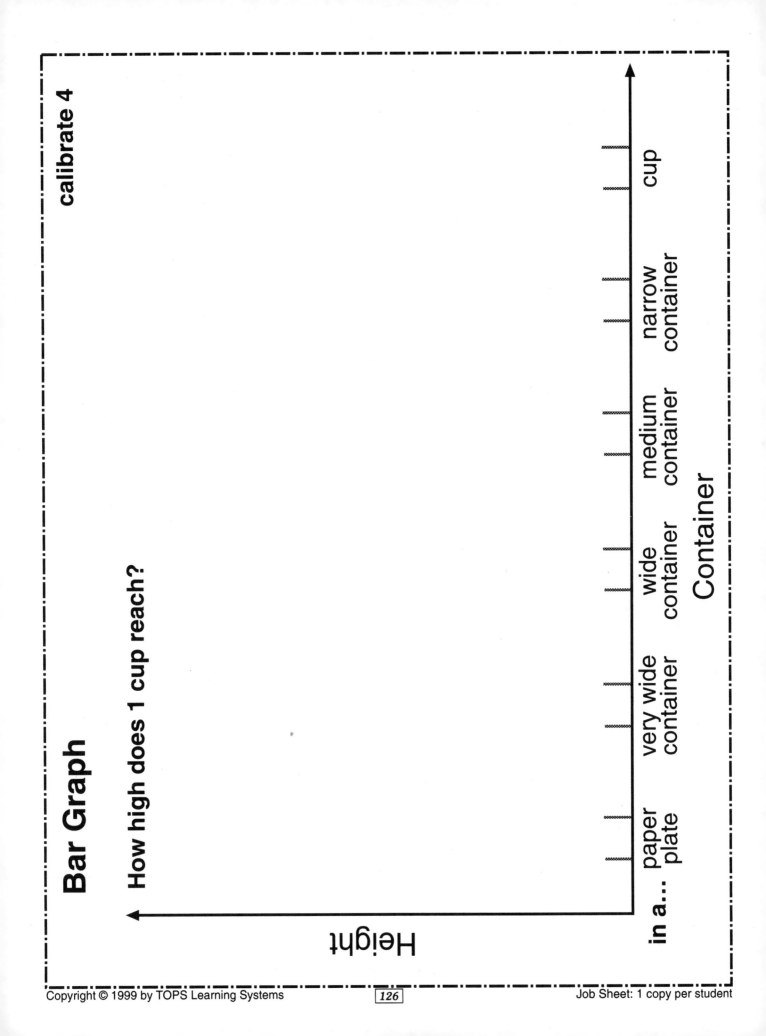

Bar Graph

How high does 1 cup reach?

calibrate 4

Height

in a...

paper plate · very wide container · wide container · medium container · narrow container · cup

Container

Job Sheet: 1 copy per student

TEACHING NOTES

Purpose

To calibrate and graph the rising height of lentils up a measuring stick, as equal volumes are added to small and large containers.

Introduction

▶ Pour a large baby food jar (BFJ) of lentils into the Small Milk Carton and tilt to level. Push a craft stick to the bottom of the carton, then record the height of the lentils by drawing a pencil mark on the stick. Number this mark "1".

▶ Repeat this process twice more with volunteers. This will almost fill the Small Milk Carton, and result in 3 numbered marks on the craft stick.

▶ Hold up the Job Sheet. Ask your class to watch how easy it is to turn these stick calibrations into a bar graph: Set the bottom of the craft stick on the baseline over "1 container" and draw a bar to the first pencil mark; hold the stick over "2 containers" and draw up to the second mark; over "3 containers" and draw up to the third mark. Finish this little graph by filling in the boxes: *Large BFJ* pours into *Small Milk Carton*.

▶ Fill the small milk carton with any unmeasured volume of lentils. Ask who can use the stick we just calibrated to tell (approximately) how many BFJ's of lentils are now inside? (Students should report answers such as *less than 3 BFJ's*, or *about 2½ BFJ's*, etc.)

Focus

◆ What container will you use for pouring, for receiving? How will you label the boxes on your graph with this information?

◆ Will you measure the level of the lentils with your thumb or with a pencil mark? *A pencil mark.*

◆ Will you number the craft stick? *Yes. I'll number upward, starting with 1 at the lowest mark.* (Students should use pencil so they can erase and reuse the sticks.)

Checkpoint

◆ May I see your calibrated craft stick and graph? (In theory, the calibrations should be equally spaced. In practice, lack of precision tends to make them quite uneven.)

◆ What is this graph telling you?

◆ Pour an unmeasured volume of lentils into the milk carton. Use your calibrated stick to measure how much is there (approximately).

◆ Can you make another bar graph using different pouring and receiving containers? How might your bar graph look different than this one?

◆ Please erase the craft sticks when you finish.

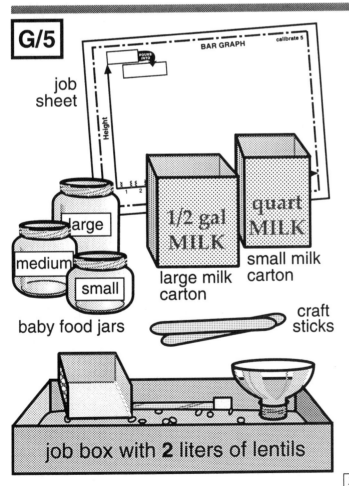

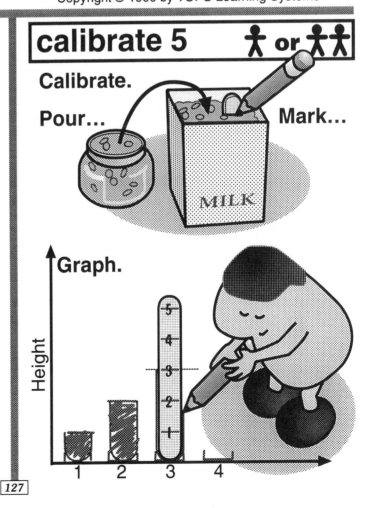

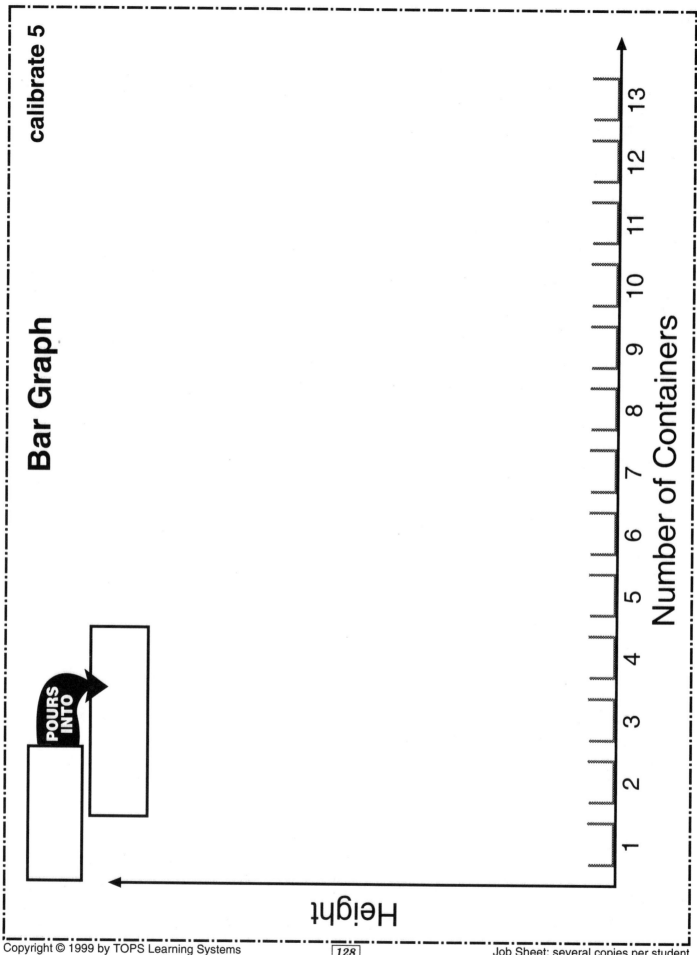

Bar Graph

calibrate 5

Height

Number of Containers

1 2 3 4 5 6 7 8 9 10 11 12 13

POURS INTO

Job Sheet: several copies per student

TEACHING NOTES

Purpose

To calibrate a tub in BFJ's. To express the relationship between volume and height on a simple line graph.

Introduction

◗ Use clear tape to stick a Calibrating Strip to the inside of a tub. Place the black triangle at the bottom (never the top).

◗ Add large BFJ's, fair and full with lentils, one at a time to the tub. Mark how high the lentils reach up the strip after each addition. (The tub will overflow after just 3 calibrations.)

◗ Tape a Job Sheet to your wall for all to see. Peel the Calibrating Strip from inside the tub and stick it over a grey patch on one of the graphs. Notice how the lines, the grey arrowhead, and the black triangle all match. Number your calibration marks 1, 2, 3, starting from the bottom.

◗ Prompt beginners with this repeated instruction: Put your fingers on matching numbers. Slide your left finger over, your right finger up, until they "crash." Draw a point at that spot and circle it.

◗ Connect the points with a straight line, but don't draw inside your circles. Label your graph: *Large BFJ* pours into *Tub*.

Focus

◆ Let me see you tape a Calibrating Strip inside the tub. (The triangle goes down.)

◆ Which size BFJ will you calibrate?

◆ (After calibrating): Let me see you move your calibrated strip to the Job Sheet. (The lines, arrow, and triangle should all match.)

◆ (After numbering): Let me see you graph points: Put your fingers on matching ___ (1's, 2's, 3's, etc.). Move over and up until your fingers "crash." Draw a point there, and circle it.

◆ What else will you do? *I'll draw a graph line (not through my circles) and title the graph.*

Checkpoint

◆ May I see your graph? What does it tell you?

◆ Can you make another line graph using different pouring and receiving containers? How might your graph line look different than this one?

More

Calibrate a clear cup in film cans, and graph the result. (Reasonably accurate graphing technique will produce a graph line that rises less steeply as the cup tapers outward. This creates a gently curved line. The more moderately tapered tubs do not register this effect as clearly.)

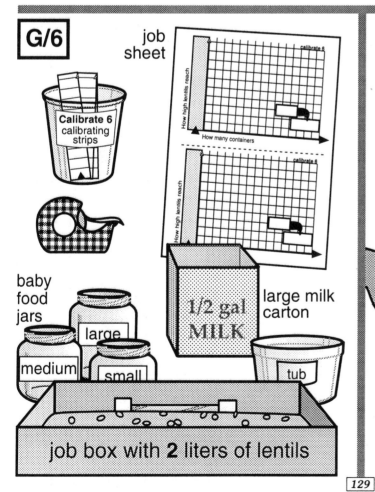

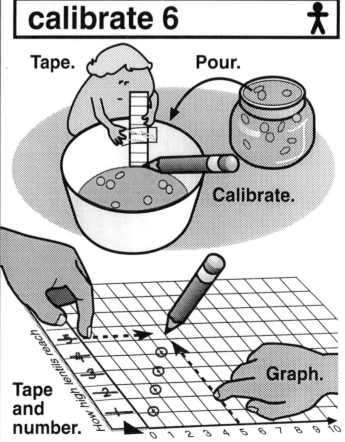

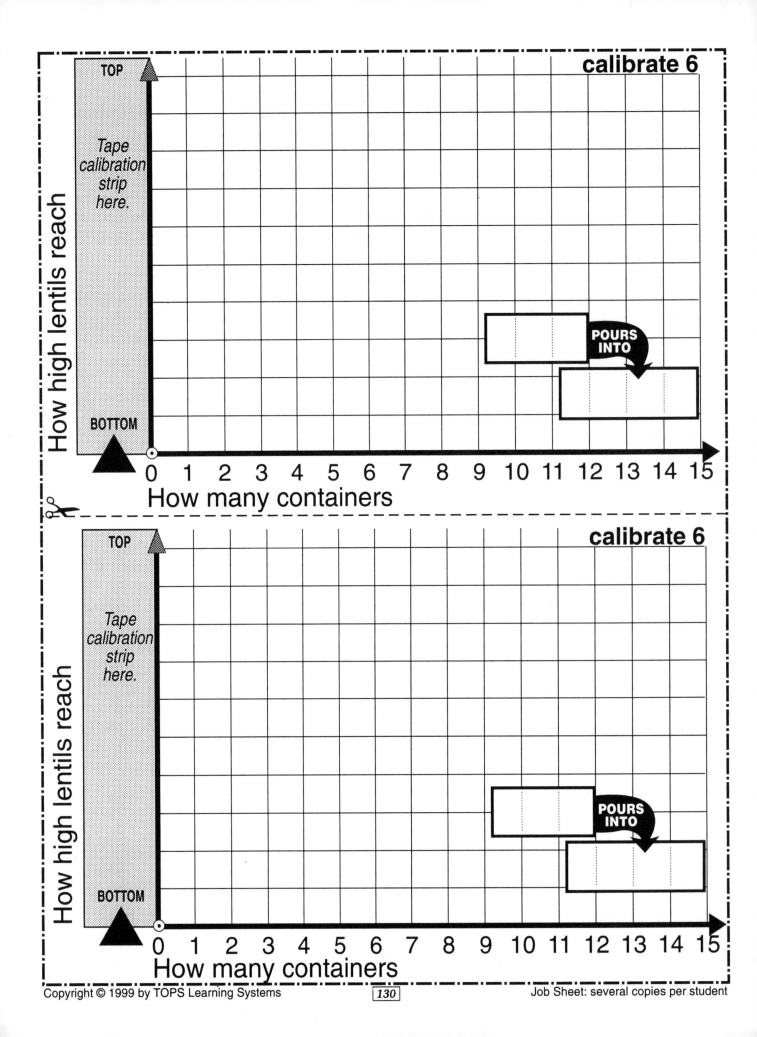

Calibrating Strips

Cut into 18 strips at the arrows.
Store in a clear cup.

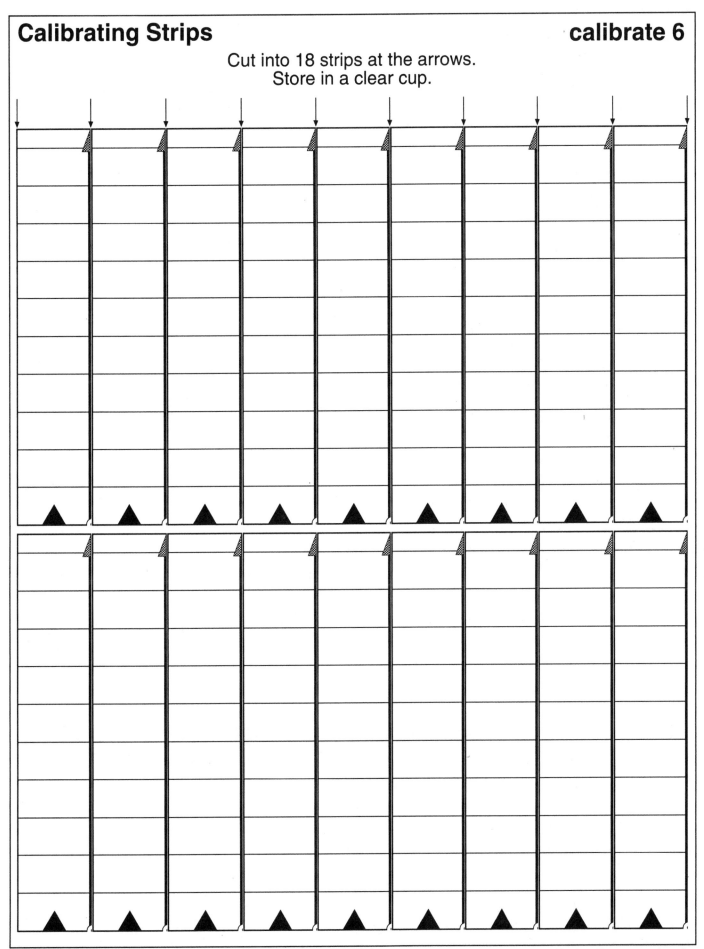

Preparation: several copies

TEACHING NOTES

Purpose

To calibrate and graph large bottles. To begin to associate the shape of the graph line with the shape of the bottle.

Introduction

♦ Watch me run a long strip of masking tape from the "neck" of this liter bottle all the way down to its "foot." Notice how I keep the tape on the roll until I cut it off at the bottom.

♦ Watch me draw a bold baseline at the bottom of this bottle. This is very important.

♦ Let's calibrate the bottle in cups. Who knows how to do this? Ask four volunteers to each funnel in 1 cup. Calibrate *and number* after each addition. Remind students to keep the lentils "fluffy" inside. If they settle, simply invert the bottle and tip to level the contents.

♦ Tape the Job Sheet up for all to see. Peel the masking tape strip from the liter bottle and stick it to the grey patch so the *baselines match*.

♦ Plot points, draw the graph line and label it in the usual manner.

♦ We can now see how the height of lentils in a liter bottle changes with the addition of each cup.

How might this graph look different when you add smaller BFJ's to this bottle instead of cups?

Focus

♦ How will you prepare the liter bottle for calibrating? *Stick masking tape down the side and draw a bold baseline at the very bottom.*

♦ Which size BFJ will you calibrate first? (More advanced students might graph all 3 sizes on the same Job Sheet. They can do this by superimposing strips of masking tape on top of each other. To distinguish data points, suggest that they surround each set of points with a different shape: circles, squares, triangles. Then add an interpretation key.)

Checkpoint

May I see your graph(s)? Why is the graph line rising?

More

Provide containers with really weird shapes to calibrate and graph. How does bottle shape affect graph line shape?

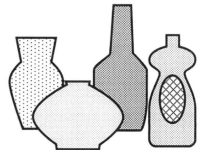

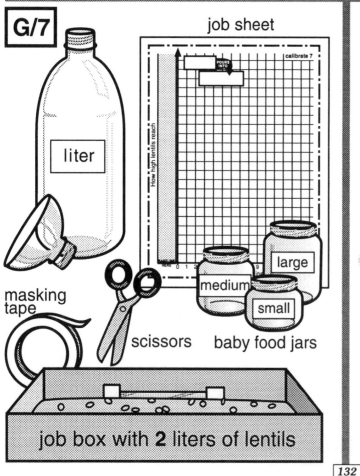

G/7   job sheet

liter

masking tape

scissors   baby food jars

How high lentils reach

POURS INTO

calibrate 7

large

medium

small

job box with **2** liters of lentils

calibrate 7   👤 or 👥

Tape, calibrate, and number.

liter

Move tape.

Graph.

1 2 3 4 5 6 7

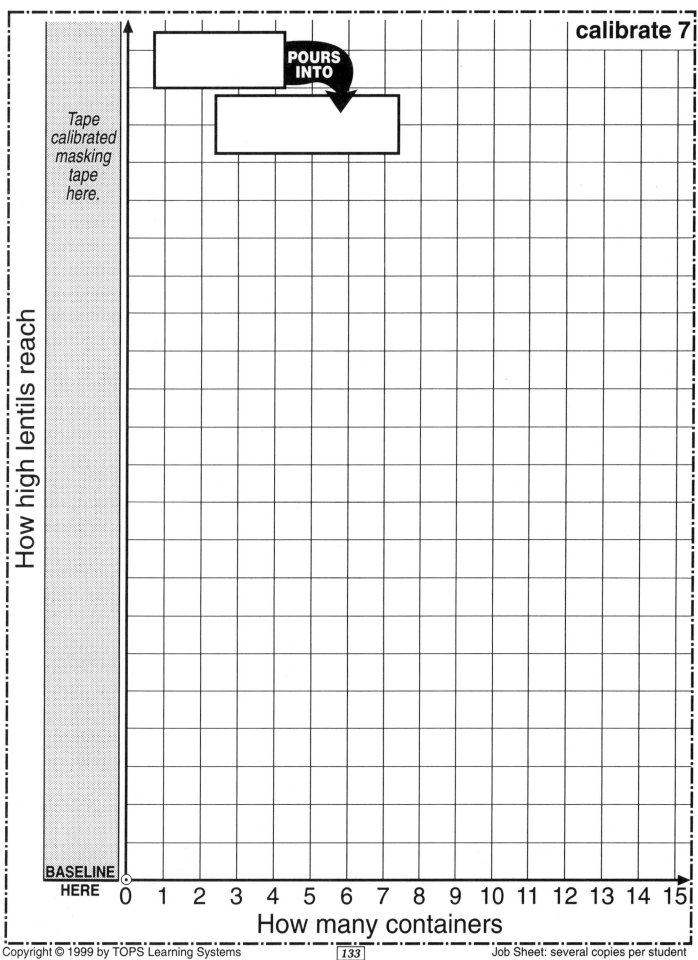

POURS INTO

How high lentils reach

Tape calibrated masking tape here.

BASELINE HERE

0 1 2 3 4 5 6 7 8 9 10 11 12 13 14 15

How many containers

Job Sheet: several copies per student

TEACHING NOTES

Purpose

To provide students with an approved way to pursue their own ideas about calibrating with lentils.

Introduction

Display this card whenever you want to do this suggested activity or experiments that you design yourself.

Focus

◆ What do you want to study about calibrating or graphing?

◆ Are these the materials you need?

Checkpoint

Report in words and pictures:
• What you did.
• What you learned.
• Questions you may still have.

Special Note

Tell TOPS about the most interesting new activities your students have designed. We may include them in future editions!

G/8 You may use...

these BASICS:

scissors

liter

small bottle

MEASURE 5 containers

masking tape

job box with **2** liters of lentils

 Ask your teacher for other items. Tell why you need them.

calibrate 8 🧍 or 🧍🧍

On your own.

My all-in-one bottle!

My idea:
Calibrate one bottle to measure everything.

qt

tub

pint

cup

½c

liter

quart

pint

cup

half cup

Feedback!

Dear Educator,

We're old hands at publishing hands-on science activities using simple things. You may already be familiar with the many books we have produced for older students. But this format and grade level are quite new to us, so your feedback would really help. Please preface your responses with the letters of the questions you are answering. We'll consider your input carefully when we reprint this book. Thanks for your time and energy!

Sincerely,

Ron & Peg

Ron and Peg Marson
author and illustrator

Primary LENTIL SCIENCE

A. Did you spot any errors, conceptual or typographical?

B. Did you find any instructions vague or misleading?

C. Were our estimates for materials useful and accurate? Did we leave anything out?

D. What grade level did you teach? Which Job Cards did you and your students like best? Least?

E. Did we strike a good balance between structure and freedom? Any tips on class management that other teachers might find useful?

F. Did you or your students think of any great Job Card ideas we should include in our next edition?

G. Any other feedback?

FOLD . . . .

TAPE CLOSED

FOLD . . . .

_____

_____

_____

PLEASE
STAMP

TOPS Learning Systems
10970 South Mulino Road
Canby OR 97013